Amateur Radio Transceiver Performance Testing

Understanding HF Transceiver Data from QST Product Reviews

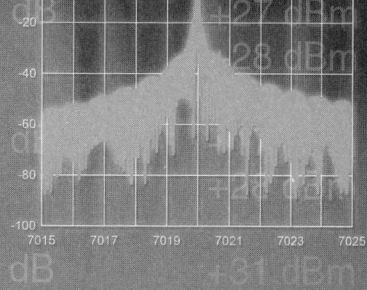

Bob Allison, WB1GCM

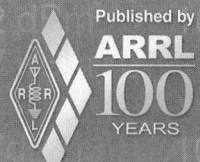

Published by
ARRL
100 YEARS

Production
Jodi Morin, KA1JPA
David Pingree, N1NAS
Shelly Bloom, WB1ENT
Maty Weinberg, KB1EIB

Cover Design
Sue Fagan, KB1OKW

Copyright © 2013 by
The American Radio Relay League, Inc.

*Copyright secured under the
Pan-American Convention*

All rights reserved. No part of this work may be reproduced in any form except by written permission of the publisher. All rights of translation are reserved.

Printed in the USA

Quedan reservados todos los derechos

ISBN: 978-1-62595-008-6

First Edition
First Printing

Please e-mail us at **pubsfdbk@arrl.org** (publications feedback) to give us your comments on this book and what you would like to see in future editions. Please include your name, call, e-mail address and the book title, edition and printing in the body of your message. Also indicate whether or not you are an ARRL member.

Contents

Foreword

Preface

About the ARRL

Section 1 — General Specifications
    1 Frequency Coverage and Modes of Operation
    2 Power Requirements

Section 2 — Receiver Performance
    3 Sensitivity (MDS) and Noise Figure
    4 Blocking Gain Compression Dynamic Range
    5 Reciprocal Mixing Dynamic Range
    6 Intermodulation Distortion Dynamic Range
    7 IF and Image Rejection
    8 AM and FM Modes
    9 Filters
    10 Receiver Audio Output
    11 Noise Reduction
    12 Other Receiver Measurements

Section 3 — Transmitter Performance
    13 Power Output, Spurious Signals and Harmonic Suppression
    14 Carrier and Unwanted Sideband Suppression
    15 Two-Tone Intermodulation Distortion Products — Transmitted IMD
    16 CW Keying Waveform
    17 CW Keying Sidebands
    18 Composite Noise Test
    19 Turnaround Times
    20 Other Considerations

Foreword

When asked about their *QST* reading preferences, over the years ARRL members have invariably listed Product Review as the most-read column each month. That's not surprising, as most hams are interested in reading about the latest gear even if they are not quite ready to step up to their next radio.

The *QST* Product Review of a modern transceiver covering the ten amateur bands from 160 through 10 meters, plus perhaps 6 meters, includes a lengthy table of performance measurements made in the ARRL Lab. All aspects of transceiver performance are checked, from the frequencies and modes covered, to the power supply requirements, to receiver sensitivity and strong signal handling, to transmitter power output and compliance with FCC requirements. Well over 100 measurements are presented in the data table for a typical HF transceiver. What does it all mean?

In this book, Senior Test Engineer Bob Allison, WB1GCM, breaks down the Product Review data table into sections. He describes how each major test is performed and explains why the test is important and what the numbers mean. Armed with that information and the descriptions and comments in the *QST* Product Review text, you'll be better able to evaluate transceivers that have caught your eye.

We hope that this book helps you to better understand the extensive testing ARRL performs on new HF transceivers under review and to answer that age-old question, "Which radio should I buy?"

David Sumner, K1ZZ
Executive Vice President
Newington, Connecticut
November 2013

Preface

The purpose of this book is to help the reader to reach his or her own conclusion as to which HF transceiver to purchase. One of the most frequently asked questions heard by the Laboratory staff is, "What radio should I buy?"

To be fair to both manufacturer and the inquisitive radio amateur, the ARRL does not give recommendations to specific products, but tries its best to "point the way" by publishing the monthly *QST* Product Review column. This helps the radio amateur to make a logical decision, based on his or her needs, including the need for the desired performance. Most Product Reviews contain one or more data tables presenting the various parameters that were measured by the ARRL Laboratory. Measurements of receiver dynamic range, sensitivity, filter characteristics and other parameters are very important to the prospective purchaser, along with transmitter output power and signal cleanliness. Each product review presents a mountain of information, and the figures and terms seen in a Product Review data table can be overwhelming.

Clearly, choosing a transceiver, whether new or used, is one of the most important decisions a radio amateur has to make. Due to the number of times I've been asked the question "What radio should I buy?," I decided it would make a good topic at forums associated with conventions at various locations around the country. Part of my work at the ARRL is public speaking, which has given me the opportunity and pleasure of meeting many of you.

During my first speaking engagement on this topic, I noticed the glazing over of eyes; some attendees interrupted my presentation with snorts of slumber. I soon realized the majority of my audience had never been exposed to the material before and I needed to stick to the basics. Granted, the definition of each specific performance term isn't exciting or easy to understand! After each engagement, many of my fellow hams encouraged me to write a book explaining, *in layman's terms*, what defines performance and how the published laboratory data relates to actual performance. After reading this book, I hope that the reader will have a better understanding of the technical terms presented and will be able to answer his or her question of, "What radio should I buy?"

ARRL members have access to all past Product Reviews from 1980 to present on our website at **www.arrl.org/product-review**. If you're interested in delving into the specific Product Review test procedures, check out the *ARRL Lab Test Procedure Manual*, available for download from the Product Review web page. And for more information on test equipment and procedures, I recommend the excellent *ARRL Handbook* "Test Equipment and Measurements" chapter by Alan Bloom, N1AL (a source for some of the information in this book).

I also encourage you to check out other websites, such as Rob Sherwood's technical data (**www.sherweng.com**) and the written words of radio amateurs found on some of the online forums and user groups.

Bob Allison, WB1GCM
ARRL Laboratory Senior Test Engineer
November 2013

Dedication

To my mother, Dena, who convinced me
to become a radio amateur.

About the ARRL

The seed for Amateur Radio was planted in the 1890s, when Guglielmo Marconi began his experiments in wireless telegraphy. Soon he was joined by dozens, then hundreds, of others who were enthusiastic about sending and receiving messages through the air—some with a commercial interest, but others solely out of a love for this new communications medium. The United States government began licensing Amateur Radio operators in 1912.

By 1914, there were thousands of Amateur Radio operators—hams—in the United States. Hiram Percy Maxim, a leading Hartford, Connecticut inventor and industrialist, saw the need for an organization to band together this fledgling group of radio experimenters. In May 1914 he founded the American Radio Relay League (ARRL) to meet that need.

Today ARRL, with approximately ~~155,000~~ members, is the largest organization of radio amateurs in the United States. The ARRL is a not-for-profit organization that:

- promotes interest in Amateur Radio communications and experimentation
- represents US radio amateurs in legislative matters, and
- maintains fraternalism and a high standard of conduct among Amateur Radio operators.

At ARRL headquarters in the Hartford suburb of Newington, the staff helps serve the needs of members. ARRL is also International Secretariat for the International Amateur Radio Union, which is made up of similar societies in 150 countries around the world.

ARRL publishes the monthly journal *QST* and an interactive digital version of *QST*, as well as newsletters and many publications covering all aspects of Amateur Radio. Its headquarters station, W1AW, transmits bulletins of interest to radio amateurs and Morse code practice sessions. The ARRL also coordinates an extensive field organization, which includes volunteers who provide technical information and other support services for radio amateurs as well as communications for public-service activities. In addition, ARRL represents US amateurs with the Federal Communications Commission and other government agencies in the US and abroad.

Membership in ARRL means much more than receiving *QST* each month. In addition to the services already described, ARRL offers membership services on a personal level, such as the Technical Information Service—where members can get answers by phone, email or the ARRL website, to all their technical and operating questions.

Full ARRL membership (available only to licensed radio amateurs) gives you a voice in how the affairs of the organization are governed. ARRL policy is set by a Board of Directors (one from each of 15 Divisions). Each year, one-third of the ARRL Board of Directors stands for election by the full members they represent. The day-to-day operation of ARRL HQ is managed by an Executive Vice President and his staff.

No matter what aspect of Amateur Radio attracts you, ARRL membership is relevant and important. There would be no Amateur Radio as we know it today were it not for the ARRL. We would be happy to welcome you as a member! (An Amateur Radio license is not required for Associate Membership.) For more information about ARRL and answers to any questions you may have about Amateur Radio, write or call:

ARRL—the national association for Amateur Radio®
225 Main Street
Newington CT 06111-1494
Voice: 860-594-0200
Fax: 860-594-0259
E-mail: **hq@arrl.org**
Internet: **www.arrl.org**

Prospective new amateurs call (toll-free):
800-32-NEW HAM (800-326-3942)
You can also contact us via e-mail at **newham@arrl.org** or check out the ARRL website at **www.arrl.org**

Section 1

General Specifications

For each transceiver reviewed in *QST*, the ARRL Lab spends many hours measuring performance, following a standard series of tests. Results are reported in a data table published in *QST*, along with the reviewer's description of the transceiver's features and functions and a subjective evaluation of how well it all works.

Throughout this book we will consider each measure of performance as presented in a typical *QST* Product Review data table. Each chapter takes a section of the table and discusses test procedures and what the results mean. The information presented in this book is current as of late 2013, although the test procedures and parameters reported evolve as technology changes.

Each table begins with a section of general specifications, including transmit and receive frequency coverage, mode(s) of operation and power supply requirements. **Table S1.1** shows an example of the general specifications for a typical HF/6 meter transceiver that operates from an external dc power supply.

Table S.1
General Specifications from a Typical ARRL *QST* Product Review Data Table for an HF/6 Meter Transceiver

Manufacturer's Specifications	Measured in the ARRL Lab
Frequency coverage: Receive, 0.1-54 MHz (specified performance, amateur bands only); transmit, 1.8-54 MHz (amateur bands only).	Receive and transmit, as specified.
Modes of operation: SSB, CW, AM, FM, RTTY.	As specified.
Power requirement: Receive, 1.8 A (no signal), 2.1 A (signal present); transmit, 23 A (100 W) at 13.8 V dc ±10%.	At 13.8 V dc: Receive, 1.88 A (VFO and backlights max brightness, max vol, no signal), 1.83 A (min brightness). Transmit, 9 A at 5 W RF output, 19 A (typical) at 100 W RF output. Operation confirmed at 12.4 V dc.

Chapter 1

Frequency Coverage and Modes of Operation

The Product Review data table begins with the transmit and receive frequencies the transceiver is capable of covering, as well as modes of operation.

Manufacturer's Specifications
Frequency coverage: Receive, 0.1-54 MHz (specified performance, amateur bands only); transmit, 1.8-54 MHz (amateur bands only).
Modes of operation: SSB, CW, AM, FM, RTTY.

Measured in the ARRL Lab
Receive and transmit, as specified.

As specified.

1.1 Receive Frequency Coverage

Frequency coverage, while not a measure of a transceiver's performance, is an important consideration when choosing an HF transceiver. Most modern HF transceivers have general coverage capability, typically extending from 100 kHz to 54 MHz. During the 1980s, manufacturers started offering solid-state transceivers with general coverage down to 500 kHz; in the 1990s, that went down to 100 kHz. At that time, longwave broadcast stations in ITU Regions 2 and 3 were still numerous between 150 kHz and 270 kHz, with channels spaced every 9 kHz. Nondirectional aero beacons, navigational sea beacons, and telegraphic traffic from ship-to-shore stations could still be heard at and below 500 kHz. Today, radio amateurs in many countries enjoy two new amateur bands: 136-138 kHz and 472-479 kHz. If you wish to

Figure 1.1 — Elecraft's K3 is a current example of a transceiver with coverage from 160 through 6 meters right out of the box.

Figure 1.2 — Icom's IC-7100 all band transceiver covers 160 through 2 meters and 70 cm. It's compact enough to work for home, mobile or portable stations.

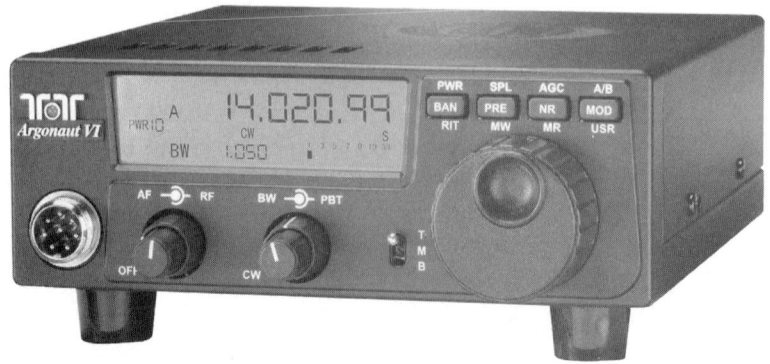

Figure 1.3 — The TEN-TEC Argonaut VI ham band-only transceiver.

receive these new ham bands, make sure the transceiver you are considering has this frequency coverage.

On the other end of the spectrum, during the late 1990s manufacturers started to include 6 meters for both transmitting and receiving. Prior to that, a separate monoband 6 meter transceiver or multiband VHF/UHF transceiver was needed. **Figure 1.1** shows a typical HF/6 meter transceiver.

A few models of so-called "all band" transceivers not only include HF and 6 meters, but also the 2 meter and 70 cm ham bands. Some even have an optional module for the 23 cm (1200 MHz) band. Many all-band transceivers are small, making them suitable for mobile or portable operation. They are capable of 100 W RF power output on the HF and 6 meter bands, 50-100 W on 2 meters, and 20-50 W or more on 70 cm. In addition to the amateur bands, all-band transceivers have receive coverage that includes commercial, aircraft, FM broadcast and other radio services. **Figure 1.2** shows a recent example of an all-band transceiver.

Some modern HF transceivers and most older HF transceivers manufactured in the mid-1980s and earlier may cover the HF amateur bands only. If you want to listen to non-amateur services, such as medium wave and shortwave stations, be sure to check out the specified frequency coverage. **Figure 1.3** shows a modern HF transceiver with ham band-only receive coverage.

1.2 Transmit Frequency Coverage

As radio amateurs, we must be certain our transmissions are made on frequencies allocated for the Amateur Service. Many HF transceivers transmit only on the amateur frequencies, but a few have been noted to transmit outside the ham bands. There are a few makes and models of HF transceivers that transmit a few kHz above and below some of the amateur bands, on frequencies intended for Military Auxiliary Radio Service (MARS). The user of such a transceiver must be especially careful to observe the band edges and to avoid transmitting outside the amateur bands.

Looking again at the Product Review data table segment at the beginning of this chapter, we can see that our sample transceiver transmits only on the ham bands. It can, however, transmit across the entire 60 meter ham band, not just on each of the specific channelized frequencies that are allocated in the United States. In this case, we must be careful to observe the transmitted frequency coverage. With such a transceiver, it is best to store the channelized frequencies in individual memory channels, if the transceiver is so equipped.

As a side note, almost all the new, inexpensive VHF/UHF handheld transceivers marketed for services other than the Amateur Service will

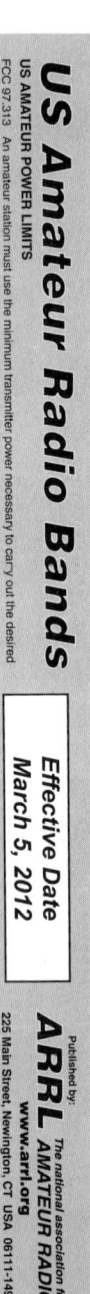

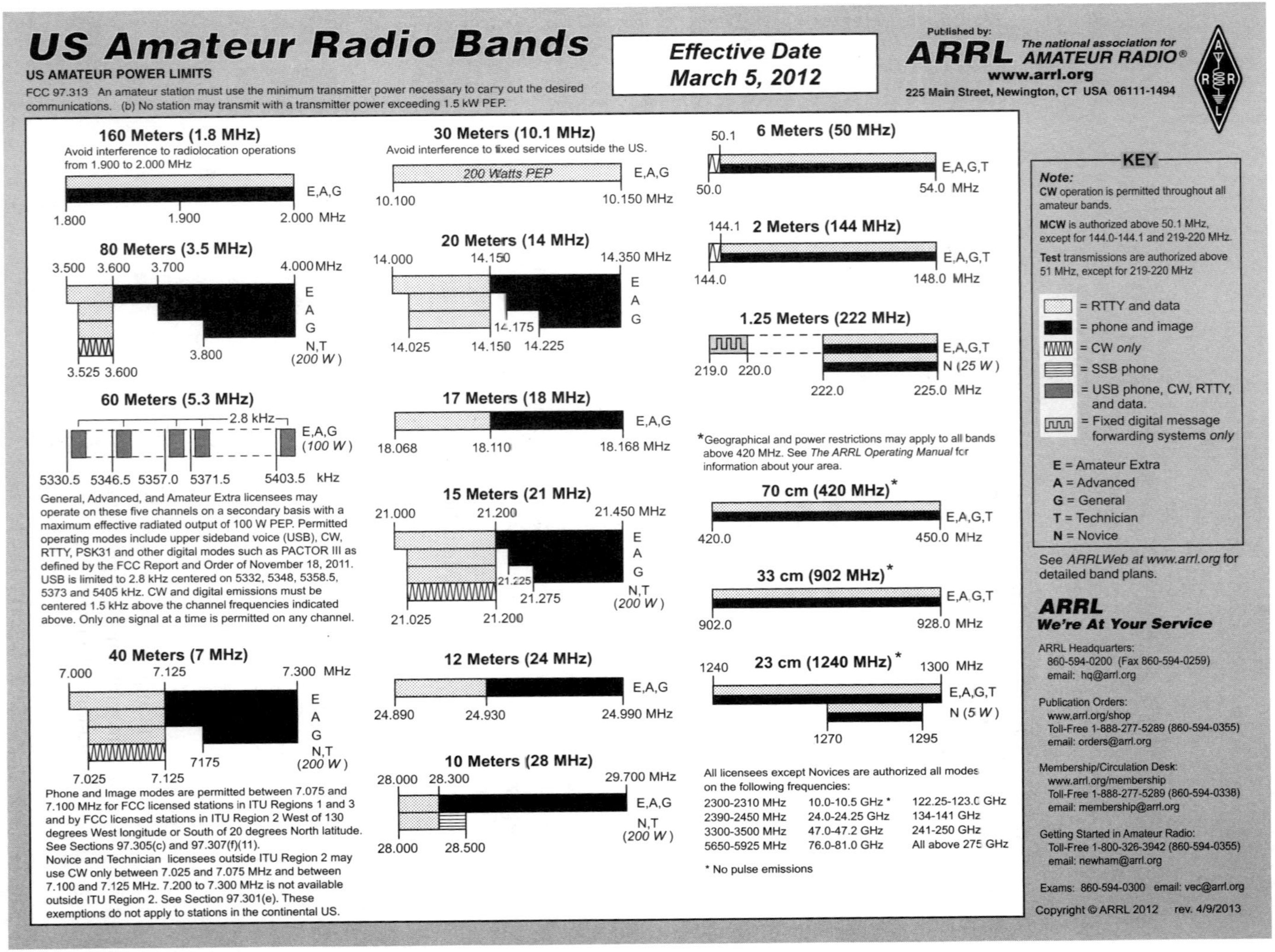

Figure 1.4 — US amateur bands.

transmit typically from 136-174 MHz and from 400-470 MHz. Our *QST* Product Review data tables for such transceivers will show the actual measured transmitter coverage. The radio amateur must be vigilant to ensure that transmissions are made only on the allocated amateur frequencies. **Figure 1.4** shows the current US amateur allocations by license class.

1.3 Modes of Operation

Current HF transceivers all include SSB and CW operation. Most also include some combination of AM, FM and digital modes (labeled RTTY, FSK, AFSK and so on). The Lab confirms operation on each mode, and there are some receiver and transmitter tests that are mode specific. Those will be covered in later chapters.

Chapter 2

Power Requirements

Power requirements show what kind of power source the transceiver requires (ac mains for an internal power supply, 13.8 V dc external supply, battery type for handhelds, and so forth).

Manufacturer's Specifications
Power requirement: Receive, 1.8 A (no signal), 2.1 A (signal present); transmit, 23 A (100 W) at 13.8 V dc ±10%.

Measured in the ARRL Lab
At 13.8 V dc: Receive, 1.88 A (VFO and backlights max brightness, max vol, no signal), 1.83 A (min brightness). Transmit, 9 A at 5 W RF output, 19 A (typical) at 100 W RF output. Operation confirmed at 12.4 V dc.

2.1 Choosing a Power Supply

Choosing an amateur HF transceiver often involves giving consideration to the necessary power supply. Is the power supply built in? If not, what type of external power supply will you need?

Today's typical transceiver with 100 W or less of RF output requires a power supply that provides enough voltage and current for it to operate as specified. Take note of the power requirements needed from a dc power supply when deciding on an HF transceiver.

For a 100 W transceiver, ARRL Lab measurements have shown typical key down operating currents of between 16 and 19 A at 13.8 V dc, but current as high as 21 A has been observed. A 20 A power supply will power a 100 W transceiver in most cases, but having a little headroom, so to speak, is good insurance for proper operation. It is wise to use a power supply that can provide at least 25 A peak. The extra capacity may come in handy as you add station accessories that require 13.8 V.

If you need to run your HF transceiver with a dc power source, note that there are two types to consider: a linear supply or a switch-mode supply (see **Figures 2.1** and **2.2**). A linear supply uses a heavy iron-core power transformer to step down the ac line voltage to a lower voltage that is then rectified, filtered and regulated. The advantage of this type is that there is little or no RF emission, which is desirable, especially if low frequency reception is needed. The disadvantage is the associated weight — 20 pounds or more for a 20 A supply.

A switch-mode supply does not use a heavy power transformer,

Figure 2.1 — Switch-mode power supplies, such as this MFJ-4125, are small and light.

Figure 2.2 — The Astron RS-20M linear power supply.

but instead incorporates somewhat complicated circuitry to produce the necessary 13.8 V dc and current. The advantage of a switch mode supply is that it is lightweight and efficient, but its disadvantage is associated RF emission, which can have a serious impact on reception at and below the AM broadcast band. The ARRL has published a number of power supply Product Reviews, each containing data tables regarding the RF emission from the power supplies reviewed.

2.2 Power Conservation

Most transceivers that require external power supplies are specified to operate over a range of voltages, typically 13.8 V dc ± 10% (12.4 to 15.2 V) or ± 15% (11.75 to 15.9 V). Most ARRL Lab testing is carried out at 13.8 V dc, but the Lab also checks operation at the minimum specified operating voltage. The transmitter power output is measured and the transceiver is monitored for proper operation while transmitting in both the continuous wave (CW) and single sideband (SSB) modes. In this case, the transceiver operated normally at 12.4 V dc, the minimum specified. If power output had been lower than the specified 100 W, that would have been noted.

It is especially important to know the minimum specified voltage when operating with battery power because battery voltage drops over time. Once the supplied voltage is lower than the minimum operating voltage, the transceiver's frequency stability and linearity can be compromised.

In some cases, limited battery power is available. Those who pay careful attention to their milliamps will take note of the current drain with the "lights off." Though it may seem frivolous to some to worry about how much the dial/display lights consume, it is critical for those who are operating under disaster or emergency conditions, and any method of conserving power is desirable. Looking at the transceiver Product Review data table segment at the top of this chapter you will see that there is a small current savings with the display set to minimum brightness.

A battery can last a long time with an HF transceiver if you turn down the 100 W transmit power when not needed to maintain contact. I have operated ARRL Field Day using a freshly charged lawn tractor battery with an HF transceiver operating at an RF power output of 5 W. The battery still started the tractor after Field Day was over.

2.3 AC Power Consumption

Some HF transceivers include an ac power supply. Transceivers in the 200 W RF output class almost always have a built-in supply or require a special external supply and are not good candidates for mobile use or portable operation with batteries.

Note that 200 W class radios can consume 700 W peak or more from a wall outlet. While a typical house circuit can handle this amount of power, the owner of a 200 W transceiver must carefully take into account all other electrical devices that use the same house circuit. If a circuit is already loaded down with other devices, such as a personal computer, monitor and lighting, voltage drops can occur on the circuit on voice peaks or with key-down conditions. This effect is most noticeable when operating at full power with an incandescent lamp plugged into the same circuit as the transceiver. If the lamp dims noticeably on voice peaks, you should consider having a separate ac circuit with circuit breaker to power the transceiver.

Here is an example of how the Product Review data table shows how a 200 W transceiver consumes power at the outlet.

Power consumption at 117 V ac: receive, no signal, 70 VA; signal present, 80 VA, transmit, 200 W output, 720 VA.

Receive, no signal, 61 VA; receive signal present, max audio, 66 VA; transmit, 481 VA at 200 W RF output.

In this case, the transceiver power consumption is rated in VA (Volt-Amperes, $V_{RMS} \times A$). The Lab measurements indicate that the radio needs 4.1 A at 117 V ac (481 VA/117 V = 4.1 A). Before purchasing an ac-powered transceiver, such as the one shown in **Figure 2.3**, determine if your ac house wiring is adequate for your needs.

Figure 2.3 — Yaesu's FTDX5000 is an example of an ac-powered, high performance transceiver with 200 W RF output.

Section 2

Receiver Performance

One hundred years ago, most radio operators used a piece of rock (galena crystal) for the detection of spark transmissions emanating from ships at sea, commercial wireless stations on shore, and Amateur Radio stations operating above 1500 kHz. Radio signals were mostly detected without electricity and were heard weakly through an uncomfortable pair of headphones. For the radio amateur, all apparatus was homemade. Selectivity was typically poor, and making matters worse, the spark transmitters of that era used far more spectrum space than today's Morse code (CW) transmitters (or any modern mode signal, for that matter). A carefully built inductor and capacitor placed in the antenna circuit determined the transmitter's operating frequency and bandwidth. Station performance was based on how far the apparatus could communicate.

Today's radio amateur enjoys single signal reception, exceptional sensitivity, and the ability to handle, to some extent, one or more strong adjacent signals. At the ARRL Lab, our observation is that, generally, more expensive transceivers have better receiver performance than less expensive transceivers. However, these measurements are made in a laboratory environment with test equipment at the antenna jack. With an actual antenna system placed at the antenna jack, performance can vary widely with conditions, manmade noise, atmospheric noise, and the gain of the antenna system.

For those interested in contesting or serious DX work, it is wise to invest in an antenna system that provides the best possible reception. Receiver performance is equally important since the signal voltage at the receive end of the transmission line is much higher with an antenna that has significant gain atop a tall tower, in contrast to the signal level obtained from a standard dipole at a modest height. The receiver must be capable of receiving weak signals in the presence of nearby strong signals on a tightly packed band.

For operators living in areas with antenna restrictions and using simple,

Table S2.1
Receiver Specifications from a Typical ARRL *QST* Product Review
Data Table for an HF/6 Meter Transceiver

Manufacturer's Specifications

Receiver
SSB/CW sensitivity: 2.4 kHz bandwidth,
  10 dB S+N/N: 0.5-1.8 MHz (IPO), 4.0 µV;
  1.8-30 MHz, 0.16 µV (preamp 2 on);
  50-54 MHz, 1.25 µV (preamp 2 on).

Noise figure: Not specified.
AM sensitivity: 6 kHz bandwidth, 10 dB S+N/N:
  0.5-1.8 MHz (preamp off), 28 µV; 1.8-30 MHz
  (preamp 2), 2 µV; 50-54 MHz (preamp 2),
  1 µV.

FM sensitivity: 15 kHz bandwidth, 12 dB SINAD:
  28-30 MHz (preamp 2), 0.5 µV; 50-54 MHz,
  (preamp 2), 0.35 µV
Spectral display sensitivity: Not specified.
Blocking gain compression dynamic range:
  Not specified.

Reciprocal mixing dynamic range: Not specified.

Measured in the ARRL Lab

Receiver Dynamic Testing
Noise floor (MDS), 500 Hz bandwidth,
  600 Hz roofing filter:

Preamp	Off (dBm)	1 (dBm)	2 (dBm)
0.137 MHz	−114	−125	−127
0.475 MHz	−125	−138	−140
1.0 MHz	−128	−139	−142
3.5 MHz	−127	−138	−141
14 MHz	−127	−138	−142
50 MHz	−125	−137	−141

14 MHz, preamp off/1/2: 20/9/5 dB
10 dB (S+N)/N, 1-kHz, 30% modulation,
  6 kHz bandwidth, 15 kHz roofing filter:
  1.0 MHz   2.60 µV (preamp off)
  3.8 MHz   0.55 µV (preamp 2)
  50 MHz   0.56 µV (preamp 2)
For 12 dB SINAD, preamp 2:
  29 MHz   0.23 µV
  52 MHz   0.21 µV
Preamp off/1/2: −100/−113/−120 dBm.
Blocking gain compression dynamic range,
  500 Hz bandwidth, 600 Hz roofing filter:

	20 kHz offset Preamp off/1/2	5/2 kHz offset Preamp off
3.5 MHz	137/141/134 dB	132/127 dB
14 MHz	137/142/136 dB	132/127 dB
50 MHz	135/139/133 dB	128/117 dB

20/5/2 kHz offset: 106/93/82 dB.

ARRL Lab Two-Tone IMD Testing (500 Hz bandwidth, 600 Hz roofing filter)*

Band/Preamp	Spacing	Input Level	Measured IMD Level	Measured IMD DR	Calculated IP3
3.5 MHz/Off	20 kHz	−23 dBm	−127 dBm	104 dB	+29 dBm
		−8 dBm	−97 dBm		+37 dBm
14 MHz/Off	20 kHz	−17 dBm	−127 dBm	110 dB	+38 dBm
		−6 dBm	−97 dBm		+40 dBm
		−86 dBm	0 dBm		+43 dBm
14 MHz/Pre 1	20 kHz	−28 dBm	−138 dBm	110 dB	+27 dBm
		−14 dBm	−97 dBm		+28 dBm
14 MHz/Pre 2	20 kHz	−36 dBm	−142 dBm	106 dB	+17 dBm
		−14 dBm	−97 dBm		+28 dBm

Band/Preamp	Spacing	Input Level	Measured IMD Level	Measured IMD DR	Calculated IP3
14 MHz/Off	5 kHz	−22 dBm	−127 dBm	105 dB	+31 dBm
		−9 dBm	−97 dBm		+35 dBm
		−80 dBm	0 dBm		+40 dBm
14 MHz/Off	2 kHz	−27 dBm	−127 dBm	100 dB	+23 dBm
		−17 dBm	−97 dBm		+23 dBm
		−71 dBm	0 dBm		+36 dBm
50 MHz/Off	20 kHz	−33 dBm	−125 dBm	92 dB	+13 dBm
		−7 dBm	−97 dBm		+14 dBm

Second-order intercept point: Not specified.

14 MHz, preamp off/1/2: +87/+75/+75 dBm; 50 MHz, +89/+75/+75 dBm.

DSP noise reduction: Not specified.
Notch filter depth: Not specified.

Variable, 30 dB maximum.
Manual: >70 dB; auto: >70 dB, attack time: 100 ms.

FM adjacent channel selectivity: Not specified.
FM two-tone, third-order IMD dynamic range: Not specified.

29 MHz, 86 dB; 52 MHz, 82 dB.
20 kHz offset, preamp 2:
  29 MHz, 86 dB; 52 MHz, 82 dB.
10 MHz channel spacing:
  29 MHz, 111 dB; 52 MHz, 105 dB.

S meter sensitivity: Not specified.

S9 signal at 14.2 MHz, preamp off/1/2: 94.3/24.8/9.2 µV.

Squelch sensitivity: Not specified.

At threshold: SSB (preamp off), 9.22 µV; FM, 29 MHz (preamp 2), 0.42 µV; 52 MHz (preamp 2), 0.33 µV.

Receiver audio output: 2.5 W into 4 Ω at 10% THD.
IF/audio response: Not specified.

2.6 W at 3.2% THD into 4 Ω (maximum audio). THD at 1 V RMS: 0.4%.
Range at −6 dB points, (bandwidth)[‡]:
  CW (500 Hz): 450-947 Hz (497 Hz)
  Equivalent Rectangular BW: 501 Hz
  USB (2.4 kHz): 164-2306 Hz (2142 Hz)
  LSB (2.4 kHz): 157-2295 Hz (2138 Hz)
  AM (6 kHz): 79-2696 Hz (5234 Hz).

Image rejection: 160-10 meters, >70 dB; 50-54 MHz, >60 dB.

First IF rejection, 10 MHz, 60 dB; 14 MHz, 77 dB; 50 MHz, 100 dB; image rejection, 14 MHz, 73 dB; 50 MHz, 70 dB.

*ARRL Product Review testing includes Two-Tone IMD results at several signal levels. Two-Tone, 3rd-Order Dynamic Range figures comparable to previous reviews are shown on the first line in each group. The "IP3" column is the calculated Third-Order Intercept Point. Second-order intercept points were determined using −97 dBm reference.
[‡]Default values; bandwidth and cutoff frequencies are adjustable via DSP. CW bandwidth varies with PBT and Pitch control settings.

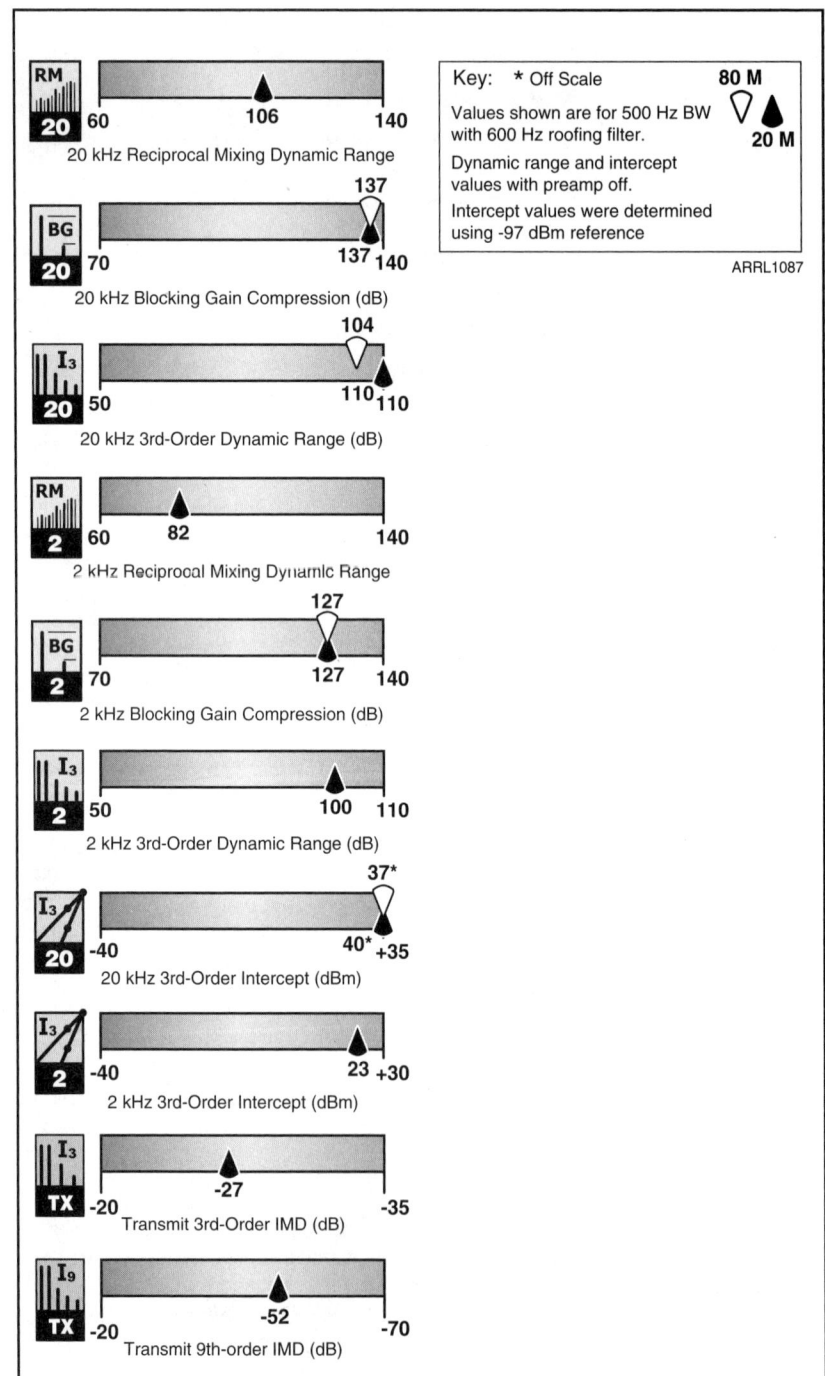

Figure I2.1 — Key Measurements Summary color chart based on the data for the example transceiver shown in Table S2.1.

low or zero gain antennas, large signal voltages are typically not present at the antenna jack. In such cases, the operator does not require top-end receiver performance. But what if the operator with a simple dipole lives near another active HF operator? In this case, some aspects of performance are important.

In the following chapters, we will consider each measure of performance as presented in a typical ARRL *QST* Product Review data table, such as that shown in **Table S2.1**. This particular HF and 6 meter transceiver operates on SSB, CW, AM and FM, and the table includes performance measurements covering all of those modes.

Readers of *QST* product reviews of HF transceivers will also recognize the Key Measurements Summary color chart (in black and white here), which provides an "at a glance" look at several of the most important measurements (**Figure S2.1**). The range of values shown below each color bar represents what has been actually observed in the ARRL Laboratory over the years for that measurement. Measurements are shown with pointers and use values from Table S2.1. Obviously, red is not so good, green is desirable, and yellow is about average.

Chapter 3

Sensitivity (MDS) and Noise Figure

How weak a signal can the receiver hear? That's the subject of these measurements.

Manufacturer's Specifications
SSB/CW sensitivity: 2.4 kHz bandwidth,
10 dB S+N/N: 0.5-1.8 MHz (IPO), 4.0 µV;
1.8-30 MHz, 0.16 µV (preamp 2 on);
50-54 MHz, 1.25 µV (preamp 2 on).

Noise figure: Not specified.

Measured in the ARRL Lab
Noise floor (MDS), 500 Hz bandwidth, 600 Hz roofing filter:

Preamp	Off (dBm)	1 (dBm)	2 (dBm)
0.137 MHz	−114	−125	−127
0.475 MHz	−125	−138	−140
1.0 MHz	−128	−139	−142
3.5 MHz	−127	−138	−141
14 MHz	−127	−138	−142
50 MHz	−125	−137	−141

14 MHz, preamp off/1/2: 20/9/5 dB

3.1 What Is MDS?

A receiver's sensitivity is the minimum signal level at which a radio signal can be heard sufficiently above the background noise to be usable. This is known as the *minimum discernible signal* (*MDS*) or *noise floor*. The MDS figure is given in dBm (decibels referenced to a milliwatt), in which 0 dBm represents a 1 mW signal into a 50 Ω load.

In a laboratory environment, the MDS is defined as a CW signal that is 3 dB above the background speaker noise (no signal applied). It is measured by placing a signal generator at the antenna jack and a relative audio meter at the speaker output (**Figure 3.1**), with the receiver's filter(s) set to a narrow bandwidth, typically 500 Hz. The table will spell out the bandwidth settings

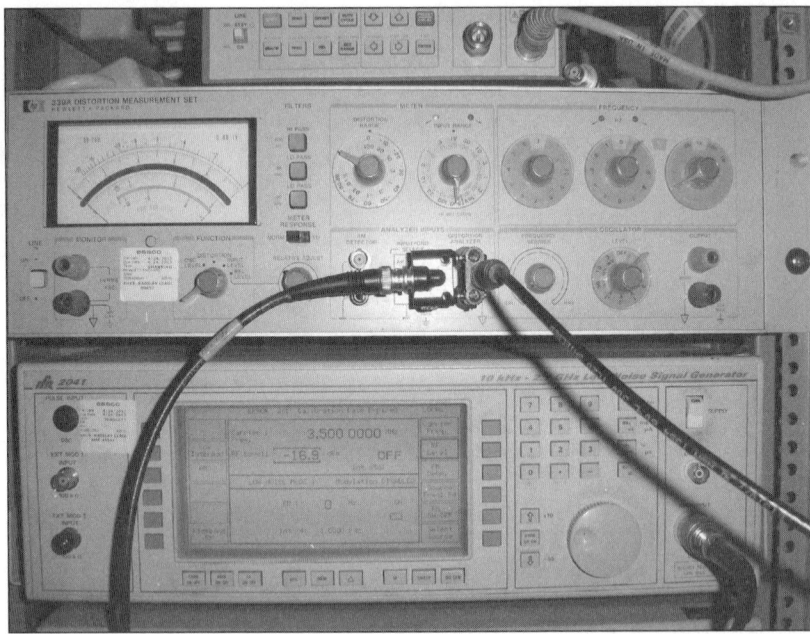

Figure 3.1 — The ARRL Lab's signal generator (bottom) and audio/distortion meter (top) are used in determining MDS level.

used for each measurement. In this case, the receiver is adjusted to use its 500 Hz DSP bandwidth setting and 600 Hz roofing filter (first IF filter). In addition, the receiver is set to the CW mode if available.

Figure 3.2 shows a test setup for determining sensitivity. The signal generator level at the input to the receiver is increased until a 3 dB increase of audio is obtained at the speaker output, as measured with the audio meter. The signal generator output level is recorded as the MDS level. The MDS level will also be used in later chapters as part of the formula for calculating the three important dynamic ranges.

While a receiver's MDS level may be as good as −135 to −140 dBm, this performance level is almost always irrelevant on the HF bands when an antenna and transmission line are placed at the antenna jack. Atmospheric, solar and manmade noise raises the background audio noise level with an antenna hooked up, and that noise will mask weak signals. In many cases, a receiver with a mediocre MDS of −120 dBm in the lab will hear actual on-air signals just as well as a very sensitive receiver with an MDS of −140 dBm.

The exception to this phenomenon is observed on 10 meters and up, where noise sources, external to the receiver's internal noise, are

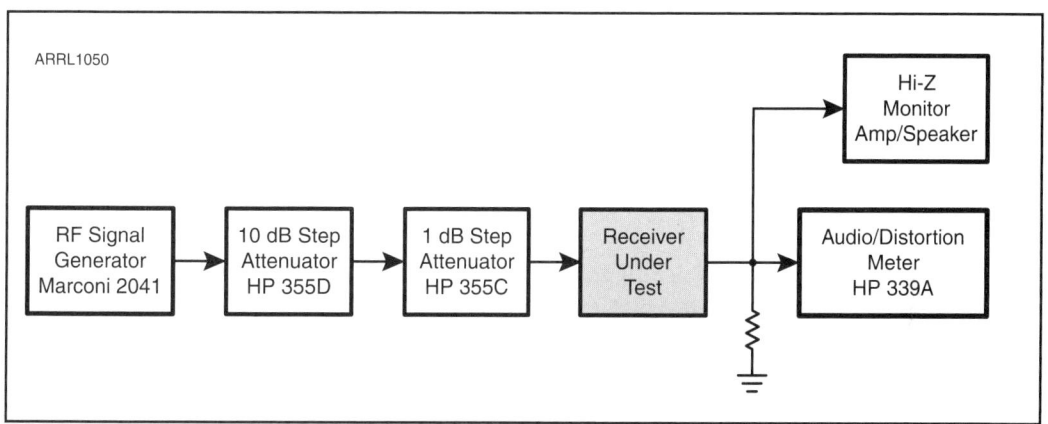

Figure 3.2 — Sensitivity (MDS) and noise figure test setup.

considerably fewer. If you enjoy the higher bands, pay close attention to the MDS measurement; the lower the MDS level, the better.

3.2 Preamps and Attenuators

Many mid- and high-performance receivers have one or more preamp settings. The MDS level of a receiver with the preamp off is always higher than it is with it on. The MDS level with the preamp off is selected by the manufacturer for optimum performance when many strong signals are present on a ham band. Preamps are for quiet conditions, such as during weak signal work on the higher ham bands. They also come in handy for operators using specialized receiving antennas or indoor "compromise" antennas. When choosing an HF transceiver, check out the preamp's effect on the MDS. The lower the number, the weaker the signal you will be able to hear during quiet band conditions on 10 and 6 meters.

Figure 3.3 — The Yaesu FT-1000MP transceiver attenuator control allows several choices for varied operating conditions.

While a preamp is an important feature for pulling out "the weak ones," another important receiver feature is the attenuator (**Figure 3.3**). Strong on-channel or adjacent signals can cause receiver overload, creating undesired effects and making listening to the desired signal difficult. Often an attenuator switched in line will reduce the negative effects from strong signals. When choosing a transceiver, read what the Product Reviews in *QST* and other sources have to say about the attenuator settings available. A transceiver with a minimum switchable 10 dB attenuator is needed for HF work. Adjustable attenuators with ranges of 6 dB to 20 dB and above are best for contesters and serious DXers.

3.3 Noise Figure

The noise figure is a measurement that quantifies the degradation of the signal-to-noise ratio caused by a receiver's internal components in its signal chain. I will not go into the math involved, but the bottom line is that the greatest sensitivity a receiver can offer, assuming a 500 Hz receiver bandwidth while operating at room temperature, is –147 dBm. Operators of digital modes can narrow the bandwidth to a sliver and detect (visually, using software) considerably weaker signals, but the receiver must have low internal noise to begin with. A noise figure of 0 dB is based on the –147 dBm level. A receiver with an MDS of –140 dBm will have a noise figure of 7 dB.

Chapter 4

Blocking Gain Compression Dynamic Range

How well can the receiver hear weak signals in the presence of strong signals? That's the subject of the next three chapters.

Manufacturer's Specifications	Measured in the ARRL Lab	
Blocking gain compression dynamic range: Not specified.	Blocking gain compression dynamic range, 500 Hz bandwidth, 600 Hz roofing filter:	
	20 kHz offset Preamp off/1/2	*5/2 kHz offset Preamp off*
3.5 MHz	137/141/134 dB	132/127 dB
14 MHz	137/142/136 dB	132/127 dB
50 MHz	135/139/133 dB	128/117 dB

4.1 What Is Blocking Gain Compression Dynamic Range?

Dynamic range is a measure of a receiver's ability to receive weak signals without being overloaded by strong signals. It is easy to design a receiver with good sensitivity to weak signals. It is also easy to design a receiver that is not overloaded by strong signals. It is much more difficult to design a receiver that can do both at the same time.

Blocking gain compression dynamic range (BGC DR) is the difference between the noise floor (MDS, see Chapter 3) and the signal level at which *blocking* occurs, that is, the level that causes a 1 dB reduction in gain for nearby weaker signals.

The effect of blocking is most noticeable when the desired signal (what you're listening to) drops in audio level, or even disappears, when a nearby

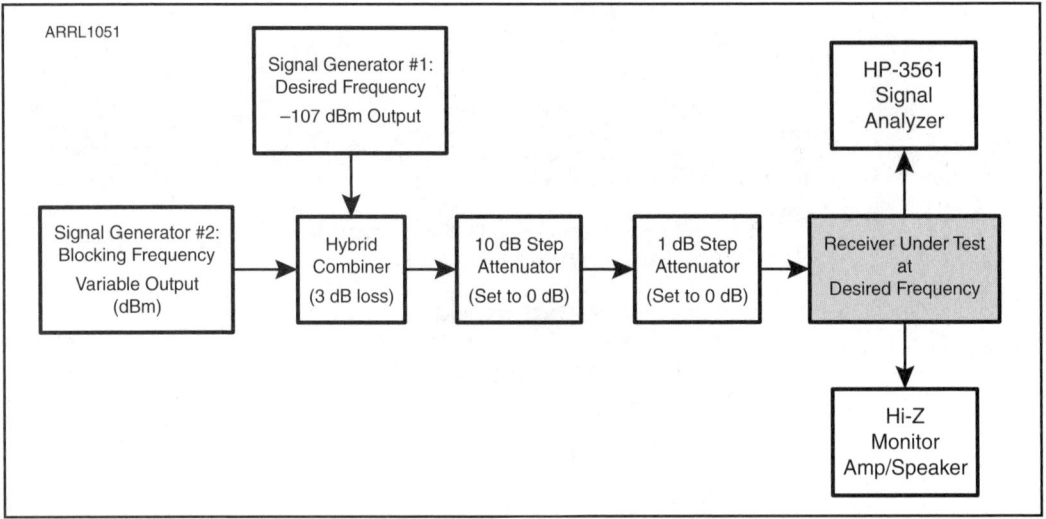

Figure 4.1 — Blocking gain compression dynamic range test block diagram. Note that the step attenuators are not used for this test.

strong signal is present. A receiver's automatic gain control (AGC) circuitry can cause the desired signal's audio level to drop, so many manufacturers offer an AGC OFF setting that allows the operator to adjust the receiver RF gain manually.

In the ARRL Lab, BGC DR measurements are made with the receiver's filter(s) set to a narrow bandwidth, typically 500 Hz. The table will spell out the bandwidth settings used for each measurement. In this case, the receiver is adjusted to use its 500 Hz DSP bandwidth setting and 600 Hz roofing filter (first IF filter), as it was for the MDS measurements in Chapter 3. The CW mode is selected and the AGC turned off.

Our BGC DR measurement is obtained with the desired signal test level well above the receiver's noise floor, with the AGC off, so that it reflects the normal signal level (linear) range of the receiver. The blocking level is measured using the test setup shown in **Figure 4.1**. Two signal generators are used. One generates the weak signal that the receiver is tuned to. The ARRL standard specifies −110 dBm at the receiver input, which requires −107 dBm (1 μV) at the input to the hybrid combiner, assuming it has 3 dB loss. The other signal generator simulates the strong interfering (blocking) signal on a nearby frequency. Standard frequency separations are ±20, 5 and 2 kHz. Note that the two step attenuators are not used (set to 0 dB attenuation) for this test.

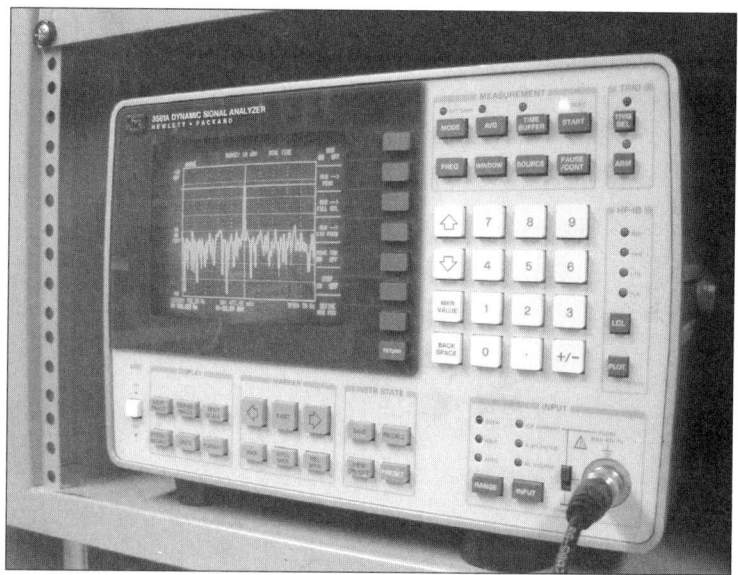

Figure 4.2 — An HP-3561A signal analyzer, used to observe a 1 dB drop of audio level.

The output level from the blocking generator is increased until a 1 dB decrease in the level of the weaker desired signal measured at the receiver audio output is observed in the audio signal analyzer (**Figure 4.2**). The input blocking level at which blocking starts is noted, and the difference between the MDS level and the blocking level determines the BGC DR. In this case, at 14 MHz with the preamp off, the MDS is −127 dBm (from Chapter 3). The blocking level (BL) is the level from the signal generator (+13 dBm) minus the loss of the hybrid combiner (3 dB):

BL = +13 dBm − 3 dB = +10 dBm

The blocking gain compression dynamic range is given by:

BGC DR = BL − MDS = +10 dBm − (−127 dBm) = 117 dB

Figures 4.3 and **4.4** show typical equipment used in blocking gain compression tests.

Figure 4.3 — A Mini-Circuits combiner is used in blocking gain compression and other tests.

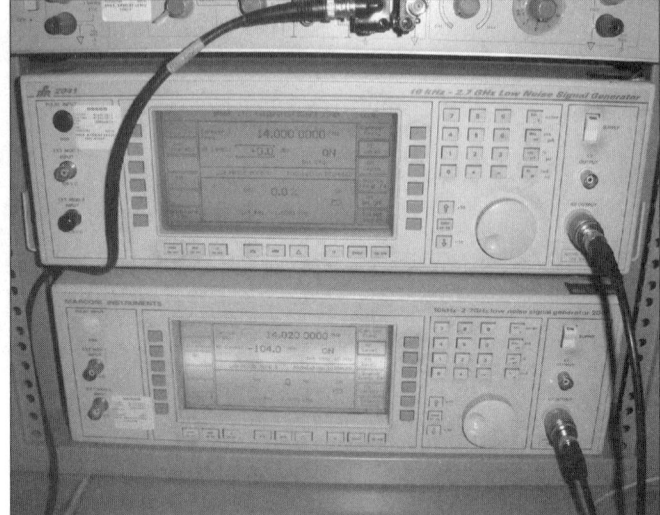

Figure 4.4 — Two signal generators are used for blocking gain compression tests.

4.2 How Much Do I Need?

Traditional analog circuitry used in most receivers can attain a BGC DR of 120 dB or higher with the blocking signal spaced 20 kHz away from the desired signal. This dynamic range decreases as the blocking signal gets closer to the desired signal, so measurements of BGC DR are made at 5 and 2 kHz spacing as well as at 20 kHz. The better the receiver's BGC DR performance, the closer the 5 kHz and 2 kHz spacing values will be to the 20 kHz values.

We have observed in the ARRL Lab that some software defined radio (SDR) receivers tend to have no blocking occur up to the point where the receiver goes into "clipping," the point at which the receiver becomes inoperable due to the analog-to-digital converter (ADC) overload.

For the serious radio amateur who regularly experiences strong adjacent signal levels, a high BGC DR is desirable. For the more modest radio amateur who perhaps lives on the same block as a Big Gun station, BGC DR is also very important. When the Big Gun is on the air, the receiver's attenuator can be switched in to reduce the input signal level and, if needed, the AGC can be disabled and the RF gain adjusted for optimum reception. For most radio amateurs with a modest antenna system consisting of dipoles or vertical antennas, excellent BGC DR may not be needed since the voltages present at the antenna connection are not high enough for blocking to occur.

Chapter 5

Reciprocal Mixing Dynamic Range

For best receiver performance, all aspects of the design must be considered. In addition to frequency stability, local oscillator performance also has a significant effect on a receiver's ability to cope with nearby strong signals.

Manufacturer's Specifications
Reciprocal mixing dynamic range: Not specified.

Measured in the ARRL Lab
20/5/2 kHz offset: 106/93/82 dB.

5.1 What Is Reciprocal Mixing Dynamic Range?

A key stage in a receiver's architecture is the first intermediate frequency (IF). This is where the input signal from the antenna mixes with the receiver's first local oscillator to produce a signal at a single (IF) frequency. This signal is then passed down the signal chain and processed until it finally reaches the audio stage.

Unfortunately, as with all oscillators, noise is generated on each side of the oscillator's carrier signal. This "sideband noise" or "phase noise" mixes with the input signal to produce noise in the speaker when a strong adjacent signal is present. *Reciprocal mixing* is the name for the mixing of a nearby interfering signal with the phase noise of the receiver's local oscillator, which causes noise in the audio output.

The ARRL Lab's reciprocal mixing test setup is shown in **Figure 5.1**. It is same as for MDS except that the signal generator is replaced with a very low-phase-noise crystal oscillator with an output power level of +15 dBm (**Figure 5.2**). (The Lab performs this test with an oscillator operating on the 14 MHz band, but any frequency could be used.) Reciprocal mixing dynamic

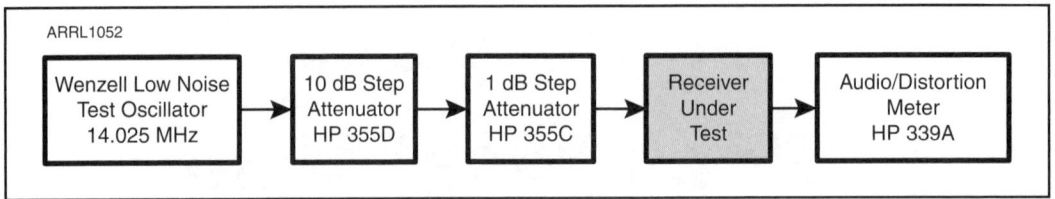

Figure 5.1 — Reciprocal mixing dynamic range test block diagram.

Figure 5.2 — Wenzel test oscillator used for reciprocal mixing test.

range is tested at 20, 5, and 2 kHz spacing, with the low-noise test oscillator simulating a strong adjacent signal.

The output noise level is first measured with the oscillator turned off. The receiver is tuned to a frequency 2 kHz (or 5 kHz or 20 kHz) away from the test oscillator frequency and adjusted to use the CW mode and the same filter settings as in the MDS test in Chapter 3. Then, using the step attenuators, the oscillator's signal level is increased until a 3 dB increase in background noise is observed with the receiver's AGC and preamp off. The signal level of the oscillator is referred to as the reciprocal mixing level. The difference between this level and the MDS level determines the reciprocal mixing dynamic range (RMDR).

In this example, the receiver has an MDS of −127 dBm (14 MHz, preamp off, from Chapter 3). The test oscillator represents a strong station operating 2 kHz away. When the test oscillator signal reaches a level of −45 dBm, it causes a 3 dB increase in noise level on the desired frequency. Reciprocal mixing equals the MDS minus the test oscillator signal strength (after attenuation), so

Reciprocal mixing = −127 dBm − (−45 dBm) = 82 dB

(Prior to 2012 we would have reported this measurement as −82 dBc, but since we now consider reciprocal mixing as a dynamic range, we report it 82 dB.)

The raised background noise created from a local oscillator's sideband noise will mask weak signals, making a receiver, at least temporarily, less sensitive. In the example receiver, if the noise floor (MDS) is −127 dBm, a signal 2 kHz away at 20 dB over S9 will cause the noise in the speaker to increase by 3 dB, and the sensitivity (MDS) of the receiver would then be reduced by 3 dB (–124 dBm). A stronger signal will create more noise, but our benchmark for testing is a 3 dB increase of noise. Operators experiencing the effects of reciprocal mixing should reduce the signal input, using the receiver's attenuator(s), if possible.

Traditional analog circuitry used in most receivers can attain a reciprocal mixing dynamic range of 100 dB or higher with the adjacent signal spaced 20 kHz away from the desired signal. (In the example receiver, it's 106 dB at 20 kHz spacing.) This dynamic range decreases as the adjacent signal gets closer to the desired signal, so measurements of RMDR are also made at 5 and 2 kHz spacing. In the ARRL Lab we have observed that SDR receivers tend to have no reciprocal mixing up to the point where the receiver goes into ADC overload.

5.2 An Important Dynamic Range

In on-air operation, RMDR can be the most limiting dynamic range of a receiver. A single strong adjacent signal can increase of background noise of a receiver and will mask the desired signal. The closer the strong adjacent signal is to the desired signal, the higher the noise level is from reciprocal mixing, since the local oscillator's sideband noise increases close to the oscillator's frequency.

Manufacturers of amateur equipment have put a lot of effort into increasing two-tone, third-order intermodulation distortion dynamic range, or 3IMD DR (we'll have more on this topic in the next chapter), where two strong adjacent signals create an undesired effect on the desired frequency the operator is tuned to. We have seen at the ARRL Laboratory that the reciprocal mixing is the most limiting dynamic range at 5 and 2 kHz spacing. A receiver with a 3IMD DR figure exceeding 100 dB from a signal 2 kHz away is meaningless if the RMDR is less than 100 dB. We hope that manufacturers will take notice of this and employ local oscillators with lower sideband noise. A serious contester or DXer with large antenna systems should check the RMDR before purchasing an HF transceiver; it is not as critical for the casual operator, unless a Big Gun is up the street.

5.3 A Note on ARRL Product Review RMDR Published Figures

The Key Measurement Summary published with each review now includes reciprocal mixing dynamic range at 20 and 2 kHz spacing. Though we have reported reciprocal mixing since December 2007, some of our technical advisors pointed out that it is easy to overlook the reciprocal mixing figures because they did not stand out from the rest of the data tables in *QST* Product Reviews of HF transceivers. Reciprocal mixing was added to the Key Measurement Summary since each of the three dynamic ranges presented must be considered to determine the overall performance of a receiver.

Chapter 6

Intermodulation Distortion Dynamic Range

Sometimes, two nearby strong signals with cause a false signal to be generated in your receiver, and the false signal can interfere with signals you are trying to hear. This chapter discusses two tests that show how well a receiver resists generating these false signals: two-tone, third-order dynamic range and two-tone, second-order dynamic range.

ARRL Lab Two-Tone IMD Testing (500 Hz bandwidth, 600 Hz roofing filter)*

Band/Preamp	Spacing	Input Level	Measured IMD Level	Measured IMD DR	Calculated IP3
3.5 MHz/Off	20 kHz	−23 dBm	−127 dBm	104 dB	+29 dBm
		−8 dBm	−97 dBm		+37 dBm
14 MHz/Off	20 kHz	−17 dBm	−127 dBm	110 dB	+38 dBm
		−6 dBm	−97 dBm		+40 dBm
		−86 dBm	0 dBm		+43 dBm
14 MHz/Pre 1	20 kHz	−28 dBm	−138 dBm	110 dB	+27 dBm
		−14 dBm	−97 dBm		+28 dBm
14 MHz/Pre 2	20 kHz	−36 dBm	−142 dBm	106 dB	+17 dBm
		−14 dBm	−97 dBm		+28 dBm

[continued on next page]

ARRL Lab Two-Tone IMD Testing (500 Hz bandwidth, 600 Hz roofing filter)* [continued]

Band/Preamp	Spacing	Input Level	Measured IMD Level	Measured IMD DR	Calculated IP3
14 MHz/Off	5 kHz	−22 dBm	−127 dBm	105 dB	+31 dBm
		−9 dBm	−97 dBm		+35 dBm
		−80 dBm	0 dBm		+40 dBm
14 MHz/Off	2 kHz	−27 dBm	−127 dBm	100 dB	+23 dBm
		−17 dBm	−97 dBm		+23 dBm
		−71 dBm	0 dBm		+36 dBm
50 MHz/Off	20 kHz	−33 dBm	−125 dBm	92 dB	+13 dBm
		−7 dBm	−97 dBm		+14 dBm

Second-order intercept point: Not specified. 14 MHz, preamp off/1/2: +87/+75/+75 dBm; 50 MHz, +89/+75/+75 dBm.

*ARRL Product Review testing includes Two-Tone IMD results at several signal levels. Two-Tone, 3rd-Order Dynamic Range figures comparable to previous reviews are shown on the first line in each group. The "IP3" column is the calculated Third-Order Intercept Point. Second-order intercept points were determined using −97 dBm reference.

6.1 What is Two-Tone, Third-Order Intermodulation Distortion Dynamic Range?

Two-tone, third-order intermodulation distortion dynamic range (3IMD DR) sounds like a complicated term, but it's actually fairly easy to understand. *Intermodulation distortion* (IMD) means the creation of spurious (undesired) signals at new frequencies because of two or more strong interfering signals modulating each other. The two-tone IMD dynamic range test is used to determine the range of signals that can be tolerated by the receiver while producing essentially no undesired spurious responses.

Two strong signals of equal spacing, say at 20 and 40 kHz above the desired frequency, combine in a receiver's front end (at the mixer) to create a false signal on the desired frequency. Our two stations at 20 and 40 kHz above the desired frequency are considered to be at 20 kHz spacing. Two stations that are 2 and 4 kHz above or below are considered to be at 2 kHz spacing, to give another example.

In a receiver's front end, the mixer combines the received input signal from the antenna jack with the local oscillator signal; the local oscillator changes frequency at the same rate as the input frequency when the tuning frequency is changed. The difference between the input signal frequency and the local oscillator frequency is the IF frequency. In the ideal world, only the difference frequency is passed from the mixer and sent to the first IF stage, but that doesn't always happen. Harmonics created in the mixer from strong input signals can also be passed along to the first IF stage (mostly second harmonics), causing undesired spurious products.

6.2 How the Signals Relate Mathematically

Let's say a receiver is tuned to a desired frequency of 14.020 MHz and that we have two strong adjacent signals appearing simultaneously at 14.040 MHz (f1) and 14.060 MHz (f2). Third-order IMD products occur at 2f1 – f2 and 2f2 – f1. In this example, the second harmonic of the 14.040 MHz signal (2f1, or 28.080 MHz) is generated in the mixer and beats with the 14.060 MHz signal. This gives us a spurious signal at 28.080 – 14.060 MHz = 14.020 MHz (the same as the desired frequency at which the receiver is tuned). See **Figure 6.1**. This spurious signal may be strong enough to interfere with the desired signal. The other IMD product is at $(2 \times 14.060) - 14.040 = 14.080$ MHz.

In the example above, a second-order signal mixes with a first-order signal, causing third-order product. Did you get all of that? That's about as difficult as this book gets.

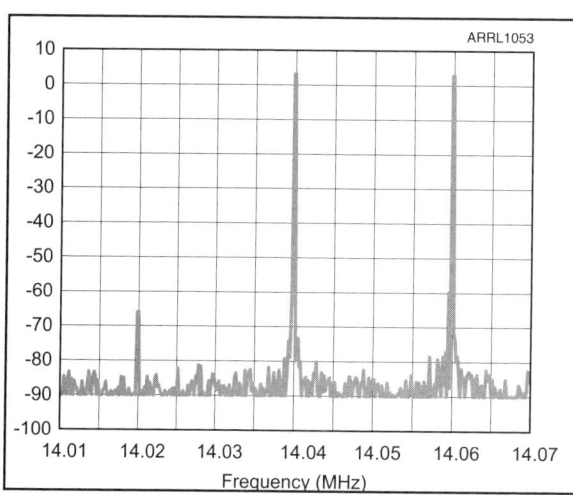

Figure 6.1 — Screen capture from an HP 8563E spectrum analyzer showing how two strong adjacent signals, spaced 20 kHz apart at 14.040 and 14.060 MHz, create an IMD signal at 14.020 MHz. Note that a spectrum analyzer is also a receiver, and like any receiver it has a 3IMD DR.

6.3 The Measurement

Figure 6.2 shows the test setup for measuring two-tone, third-order IMD dynamic range. The receiver is set to CW mode and uses the same filter settings as in the MDS test in Chapter 3.

Signal generators #1 and #2 are set to frequencies that will create an IMD signal at or very close to the desired signal. In this case, they are set for 20 kHz spacing, at 13.980 and 14.000 MHz to produce an IMD product at the frequency of the desired signal, 14.020 MHz. Additional tests are performed with generators #1 and #2 set to 14.010 and 14.015 MHz (5 kHz spacing) and 14.016 and 14.018 MHz (2 kHz spacing). The signal generators are followed

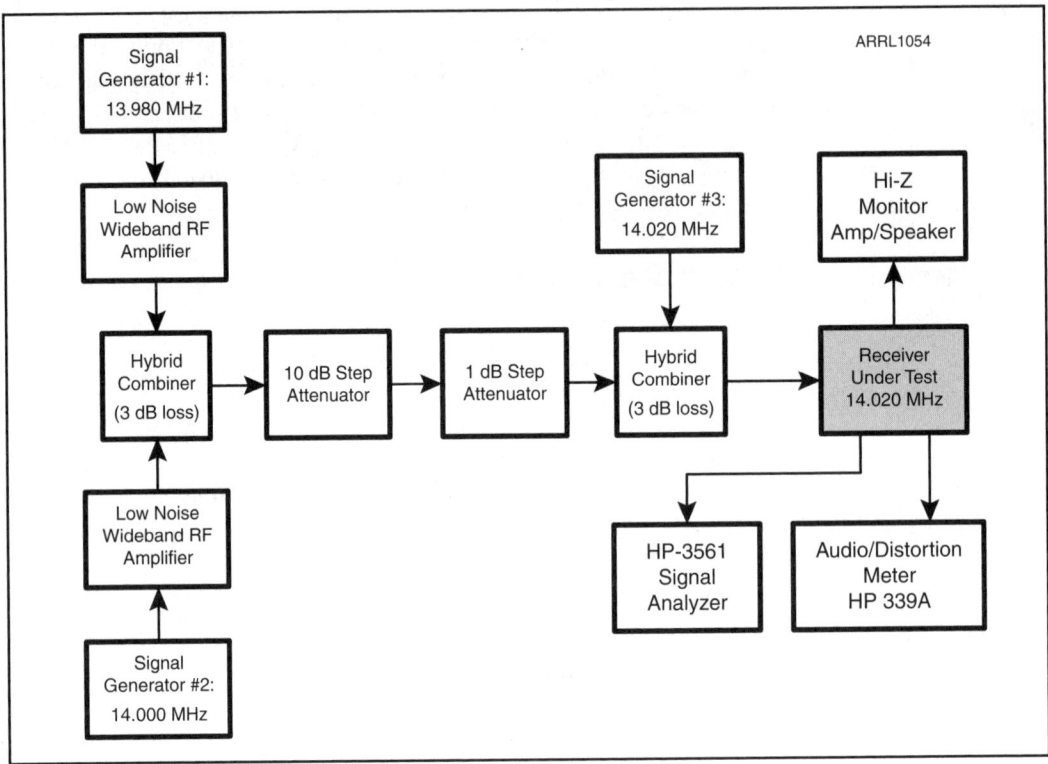

Figure 6.2 — Two-tone, third-order and second-order intermodulation distortion dynamic range test setup.

by low noise amplifiers to obtain sufficient output power and isolation between them. The outputs are then combined using a hybrid combiner, and the signal levels are adjusted so that the signal from each generator/amplifier is the same level at the combiner output as shown in **Figure 6.3**.

The combined signal from generators #1 and #2 (the IMD signal) is fed through step attenuators to another hybrid combiner, where it is mixed with the output from signal generator #3, set to the frequency of the desired signal. Next, the combination of IMD signal and desired signal is fed into the antenna jack of a receiver. The audio from the receiver's external speaker jack is then fed into a signal analyzer so we can visually observe both the IMD and desired audio signals. **Figure 6.4** shows the entire test setup.

The test is performed with the desired signal at one of three levels: equal to the MDS (as measured in Chapter 3), S5 on an S meter (–97 dBm), and 0 dBm (very strong). The step attenuators are adjusted until the IMD signal

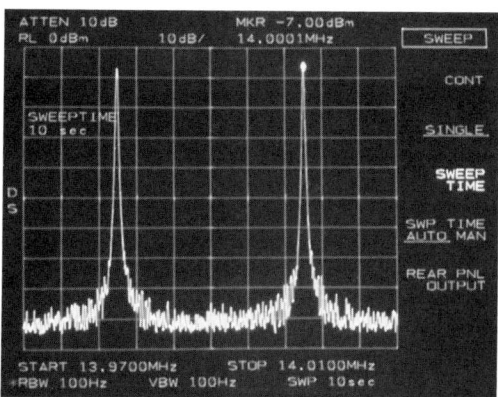

Figure 6.3 — Signal generators #1 and #2 are followed by low noise amplifiers and a hybrid combiner. The combiner output is measured with an HP 8563E spectrum analyzer and the levels adjusted until they are equal. Each level is adjusted to measure +3 dBm to make up for the 3 dB loss in the last combiner, resulting in a 0 dBm signal that feeds the receiver antenna jack prior to adjusting the step attenuators. An HP-437B power meter is then used to measure each level as an accuracy check.

level equals the desired signal level. When matched up, the IMD signal level is noted and the 3IMD DR is determined by the difference between the MDS value and the IMD signal value. In the case of the example receiver, at 14 MHz with the preamp off, at 20 kHz spacing, the MDS is –127 dBm and the IMD signal (called "input level" in the table) is –17 dBm. So:

3IMD DR = (–127 dBm) – (–17 dBm) = 110 dB

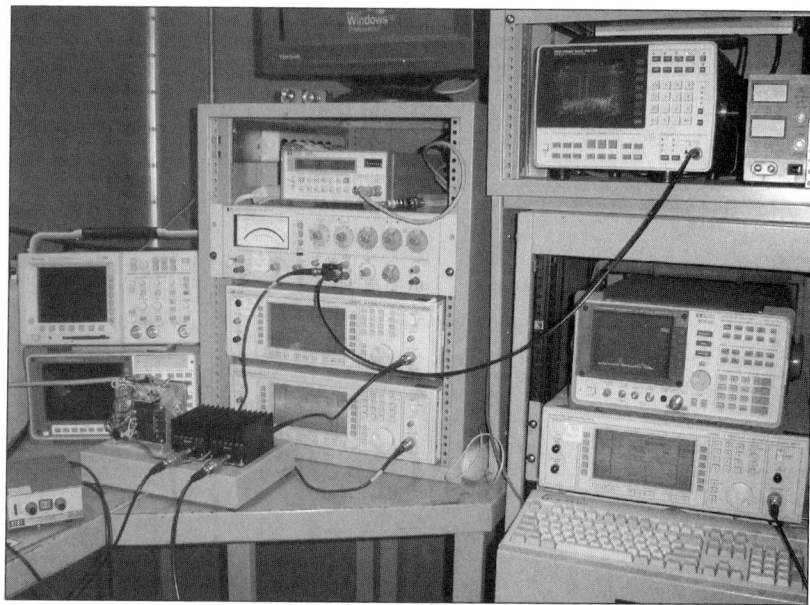

Figure 6.4 — Three signal generators are used for the two-tone intermodulation distortion dynamic range tests.

Intermodulation Distortion Dynamic Range

The test is performed at 3.5, 14 and 50 MHz with various signal spacings and with the preamp(s) on and off. In the best modern receiver designs, narrow first IF filters (roofing filters) eliminate most strong adjacent signals that could mix in the receiver and create IMD products. In the best receivers, there is little or no difference among measurements at 20, 5 and 2 kHz spacing. In older designs, and in lower cost receivers without narrow roofing filters, 3IMD DR can be significantly lower at narrow spacings.

The 3IMD DR figure is very important to the contester or DXer or low band operator with a large antenna array that delivers strong signal voltages to the antenna jack. During times when the band is crowded, multiple strong signals above and below the operating frequency can create IMD. For CW contests, IMD products may sound like an indecipherable CW signal, or like a real transmitting station if one of the two signals is tuning up. For SSB, IMD products will sound like a garble of audio products. Keep in mind that the receiver operator may experience reciprocal mixing noise (see Chapter 5) before realizing that IMD products are present, especially at 2 kHz spacing, if the reciprocal mixing dynamic range (RMDR) is lower than the 3IMD DR.

This is not the case with some SDR receivers, which experience no reciprocal mixing up to the point of analog-to-digital converter (ADC) overload. In fact, their IMD behavior is of a different stripe than traditional analog receivers. An SDR's 3IMD DR may measure as mediocre in the lab, but get considerably better when signals are present in the passband. For SDR receivers, we inject a strong fourth signal, which represents signals on the band, and measure the 3IMD DR at the MDS level. We report the 3IMD DR as "up to XX dB." For a further explanation, see October 2010 *QST,* page 52.

6.4 Third-Order Intercept Point

In most analog components such as mixers and amplifiers, the second-order products increase in amplitude by 2 dB for each 1 dB increase in the interfering signals and third-order products increase 3 dB per dB increase. If the output signal levels are plotted versus the input levels on a log-log chart (that is, in units of dB), the desired signal and the undesired IMD products theoretically trace out straight lines. Although the IMD products increase more rapidly than the desired signal, the lines never actually cross because blocking occurs before that level. The point where the extensions of those two lines cross is called the *third-order intercept point* (IP3) as shown in **Figure 6.5**.

The ARRL Lab measures the 3IMD DR at three levels to see if the receiver operates in a linear fashion, which we determine by calculating

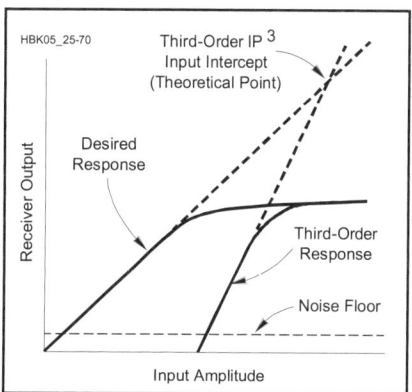

Figure 6.5 — The third-order intercept point can be determined by extending the lines representing the desired signal and the third-order intermodulation products on a plot of the signal levels in dB.

the third-order intercept point at each level. The IP3 is not a measurement, but rather a figure derived from a calculation. Although IP3 is an artificial point, it is a useful measure of the strong-signal-handling capability of a receiver. At the MDS, it can be calculated from the equation

$$IP3 = MDS + (1.5 \times 3IMD\_DR) \text{ dBm}$$

where IP3 is the third-order intercept point, MDS is the minimum discernible signal in dBm and 3IMD_DR is the third-order IMD dynamic range in dB.

Using the numbers from our example, at 14 MHz, with 20 kHz signal spacing and the preamp off:

$$IP3 = -127 + (1.5 \times 110) = +38 \text{ dBm}.$$

At the S5 (–97 dBm) level, the calculation is

$$IP3 = \frac{3 \times (S5 \text{ Input Level}) - (S5 \text{ Reference Level})}{2} = \frac{3 \times (-6 \text{ dBm}) - (-97 \text{ dBm})}{2} = +39.5 \text{ dBm}$$

At the 0 dBm level, the calculation is

$$IP3 = \frac{|\text{Input Level}| \text{ dBm}}{2} = \frac{86 \text{ dBm}}{2} = +43 \text{ dBm}$$

The third-order intercept is generally not a valid concept for software-defined receivers (SDRs) that do not use an analog front end. Some SDRs do not use a mixer but feed the signal from the antenna directly to an analog-to-digital converter (ADC). ADCs usually do not exhibit the 3 dB per dB relationship between signal level and third-order products, at least over major portions of their operating range.

6.5 Second-Order IMD DR and Second-Order Intercept Point

Two-tone, second-order dynamic range is the range of input signals at which a receiver is free from second-order products. Like third-order products, second-order products are created by the receiver's mixer and result in false signals on the dial where there shouldn't be any signals. An excellent example of this occurs when a radio amateur lives in the vicinity of strong shortwave broadcast stations, a problem for many European amateurs. You may find that a 6 MHz station and a 15 MHz station are received strongly on their assigned broadcast frequencies and add up to be a jumble at 21 MHz.

The measurement method is the same as for the 3IMD DR measurements, except the frequency pairs are very widely spaced and are picked for likely scenarios, such as the one mentioned above. For the second-order IMD measurement, the standard ARRL test sets the two signal generators to 6.000 MHz and 8.020 MHz and the receiver is tuned to 14.020 MHz. Other combinations of frequencies are checked on other HF and VHF amateur bands.

The Product Review data table reports second order intercept point (IP2) calculated at the –97 dBm (S5) reference level. At 14 MHz, preamp off:

$$IP2 = (2 \times S5 \text{ IMD level}) - (-97 \text{ dBm}) = (2 \times -5 \text{ dBm}) - (-97 \text{ dBm}) = +87 \text{ dBm}$$

The higher the IP2, the better. This data is not included as one of our Key Summary Measurements and is often overlooked. I would consider an IP2 of +30 dBm to be poor, +50 dBm to be mediocre, +60 dBm to be good, and +70 dBm to be excellent. The effects of second- and third-order products can be reduced by turning off the preamp, if used, and/or turning on the attenuator.

Chapter 7

IF and Image Rejection

The first IF and image rejection tests determine the level of signal input to the receiver at the first IF and image frequencies that will produce an audio output equal to the MDS level.

Manufacturer's Specifications
Image rejection: 160-10 meters, >70 dB; 50-54 MHz, >60 dB.

Measured in the ARRL Lab
First IF rejection, 10 MHz, 60 dB; 14 MHz, 77 dB; 50 MHz, 100 dB; image rejection, 14 MHz, 73 dB; 50 MHz, 70 dB.

7.1 IF Rejection

The frequency of the first IF (intermediate frequency) is chosen by design. Receivers that are designed for best performance on the ham bands sometimes have a first IF frequency in the 8 to 10 MHz range. If general frequency coverage outside the ham bands (say, 500 kHz through 30 MHz) is desired, the first IF frequency can be in the 45 to 60 MHz range, or higher. Needless to say, if a strong signal is received exactly on a receiver's first IF frequency, the operator will experience interference throughout the entire frequency range of the receiver. This can be a severe limitation, but it is rarely encountered since IF rejection is a major receiver design consideration.

Figure 7.1 is a block diagram of an IF and image rejection test setup. It's the same setup used in Chapter 3 to determine MDS. As with other receiver tests discussed in earlier chapters, the receiver is adjusted to use its 500 Hz DSP bandwidth setting and 600 Hz roofing filter (first IF filter). In addition, the receiver is set to the CW mode, AGC is off and the receiver is tuned to 14.020 MHz. (We always measure IF rejection on 20 meters and

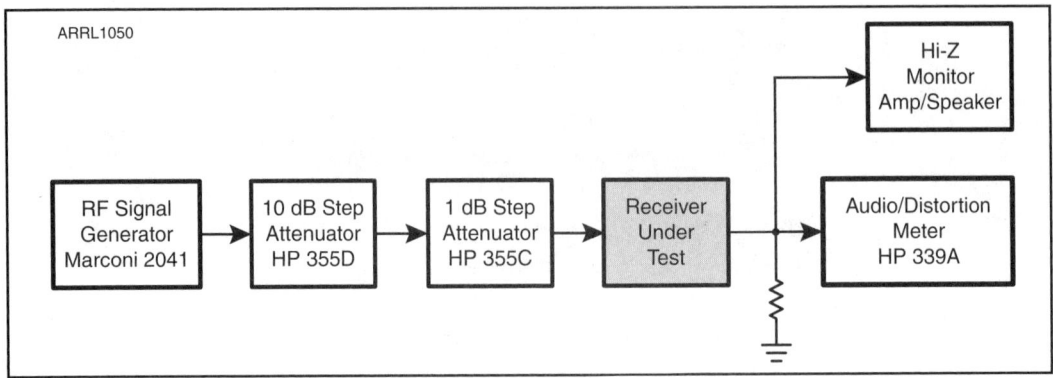

Figure 7.1 — IF and image rejection test block diagram.

An IF Prank

As a youngster participating in ARRL Field Day, I talked to an old-timer who was quite a wise-guy. He boasted that back in the 1950s, when transistors were new, he made a 455 kHz crystal oscillator that had a small Morse code key and a short length of wire attached. He concealed this mini station in his pocket, and every once in a while sent "hi" from it, which made a rival in his local ham radio club crazy since the receiver used had an IF frequency of 455 kHz. No matter where his rival tuned he would hear the word, "hi." The poor fellow never caught on. Most modern HF transceivers today would be immune to this type of prank.

sometimes extend that to other bands. In this case, IF rejection was also measured on the 10 MHz and 50 MHz bands.)

IF rejection is measured by placing a signal generator at the antenna jack, with the signal generator adjusted to the IF frequency. The signal generator output starts at a low level (–80 dBm) and the level is increased until a signal is heard in the receiver speaker that is 3 dB above the background noise as measured on the audio meter. The signal generator level is noted and recorded as the IF rejection level. The IF rejection is determined by the difference between the MDS level and the IF rejection level.

The example transceiver has a first IF frequency of 9 MHz. Our measurements determined the IF rejection to be 60 dB on the ham band closest to the IF frequency (30 meters). In this case, the receiver MDS measured –127 dBm at 10 MHz with the preamp off, and a signal at the 9 MHz IF frequency with a level of –67 dBm created a signal on the desired 30 meter frequency that was 3 dB above the background noise.

In this instance, the IF rejection is calculated from:

IF rejection = –127 dBm – (–67 dBm) = 60 dB

That means a –67 dBm signal at the 9 MHz IF created an MDS level signal on the desired frequency range of the receiver. Note that the –67 dBm signal is equivalent to S9 + 6 dB on a calibrated receiver S meter, not an unusually strong signal to run across on the air. A stronger signal at 9 MHz would create even more havoc for the operator using this receiver.

When choosing an HF transceiver, take careful note of the measured IF rejection; the higher the IF rejection, the better. An 80 dB or higher IF rejection level is desirable.

7.2 Image Rejection

Along with the first IF frequency, there are one or two frequencies where a strong signal can cause interference — the first IF image frequency. Rejection of signals at the image frequency is tested using the same setup as for IF rejection, except the signal generator is tuned to the image frequency:

Image frequency (MHz) = Desired Frequency + (2 × IF Frequency)

or

Image frequency (MHz) = Desired Frequency – (2 × IF Frequency)

Note sometimes there are two potential image frequencies, and other times there is only one. With a desired frequency of 14.020 MHz, there are two potential images only if the first IF is less than 7.010 MHz. In this case:

Image frequency = 14.020 MHz + (2 × 9.0 MHz) =
14.020 + 18.0 = 32.02 MHz

If there is a transmitter with a strong signal that falls on a specific frequency that fits this formula (32.02 MHz in this case), interference can occur. This was frequently a problem during the days of simple single conversion receivers with an IF of 455 kHz.

For the example transceiver, our measurements determined the image rejection to be 73 dB at 14.020 MHz. The receiver MDS measured –127 dBm at 14 MHz with the preamp off, and a signal at the 32.02 MHz image frequency with a level of –54 dBm created a signal on the desired 20 meter frequency that was 3 dB above the background noise.

In this instance, the image rejection is calculated from:

IF rejection = –127 dBm – (–54 dBm) = 73 dB

That means a –54 dBm signal (S9 +19 dB on the S meter) at the 32.02 MHz image frequency created an MDS level signal on the desired frequency range of the receiver. It is common to observe image rejections greater than 90 dB. The rule of thumb is: the higher the image rejection, the better.

Chapter 8

AM and FM Modes

Some HF transceivers include AM and FM voice modes in addition to SSB. These receivers are the subject of additional tests.

Manufacturer's Specifications	Measured in the ARRL Lab
AM sensitivity: 6 kHz bandwidth, 10 dB S+N/N: 0.5-1.8 MHz (preamp off), 28 µV; 1.8-30 MHz (preamp 2), 2 µV; 50-54 MHz (preamp 2), 1 µV.	10 dB (S+N)/N, 1-kHz, 30% modulation, 6 kHz bandwidth, 15 kHz roofing filter: 1.0 MHz 2.60 µV (preamp off) 3.8 MHz 0.55 µV (preamp 2) 50 MHz 0.56 µV (preamp 2)
FM sensitivity: 15 kHz bandwidth, 12 dB SINAD: 28-30 MHz (preamp 2), 0.5 µV; 50-54 MHz, (preamp 2), 0.35 µV	For 12 dB SINAD, preamp 2: 29 MHz 0.23 µV 52 MHz 0.21 µV
FM adjacent channel selectivity: Not specified. FM two-tone, third-order IMD dynamic range: Not specified.	29 MHz, 86 dB; 52 MHz, 82 dB. 20 kHz offset, preamp 2: 29 MHz, 86 dB; 52 MHz, 82 dB. 10 MHz channel spacing: 29 MHz, 111 dB; 52 MHz, 105 dB.

8.1 Adding Two Modes to the Mix

Not all modes are for contesting or serious DXing. For the radio gladiator, CW, SSB and digital are the modes of choice. The AM and FM modes are for more casual operating. AM, for instance, is used by vintage radio enthusiasts who maintain and operate receivers and transmitters that use "hollow state" (vacuum tube) technology (**Figure 8.1**), though some have designed and built modern, solid-state equipment as well. AM operation can be found on or near the AM calling frequencies.

Some transceivers include FM voice as well. In this transceiver, the FM mode is used above 29 MHz on 10 meters and on 6 meters for simplex and repeater operation.

Figure 8.1 — The author's ham shack, showing the vintage side of the room. The other side has modern HF, VHF and UHF amateur equipment.

Receiver sensitivity in the AM and FM modes is measured in microvolts (μV), with a 1 μV signal being approximately equal to a –107 dBm signal level. See **Table 8.1** for conversion between dBm and volts. For both AM and FM, the lower the microvolt figure, the better. In the ARRL Lab we have observed that some transceivers demonstrate good sensitivity in the CW and SSB modes, but mediocre sensitivity on either AM or FM, or on both. I consider sensitivities of 0.5 μV on AM and 0.25 μV on FM to be good.

Radio amateurs who wish to have AM and/or FM as a voice option will need to determine if additional filters are required. One may wish to add at least a 6 kHz IF filter for AM and a 15 kHz IF filter for FM. Some transceivers require an optional filter to activate each mode, while other transceivers include one or both modes as a standard feature.

Table 8.1
dBm to Volts Conversion

+10 dBm	0.710 V	–73 dBm	50 μV *59*
0 dBm	0.225 V	–80 dBm	225 μV
–10 dBm	0.071 V	–90 dBm	7.1 μV
–20 dBm	22.5 mV	–100 dBm	2.25 μV
–30 dBm	7.1 mV	–107 dBm	1.00 μV
–40 dBm	2.25 mV	–110 dBm	710 nV
–50 dBm	0.71 mV	–120 dBm	225 nV
–60 dBm	0.225 mV	–130 dBm	71 nV
–70 dBm	71 μV	–140 dBm	23 nV

8.2 AM Sensitivity

AM sensitivity is measured as an AM signal level that is about 10 dB above the background noise of a receiver. This AM signal also has internal receiver noise riding along with it, so the measurement is actually considered to be 10 dB signal plus noise, divided by the noise (you'll see this written as 10 dB S+N/N).

Figure 8.2 is a block diagram of an AM sensitivity test setup, which is the same setup used in Chapter 3 for the CW/SSB sensitivity measurement. The test setup requires the receiver under test to be adjusted for a 6 kHz filter bandwidth and the signal generator modulation level set to 30%, using a 1000 Hz audio tone.

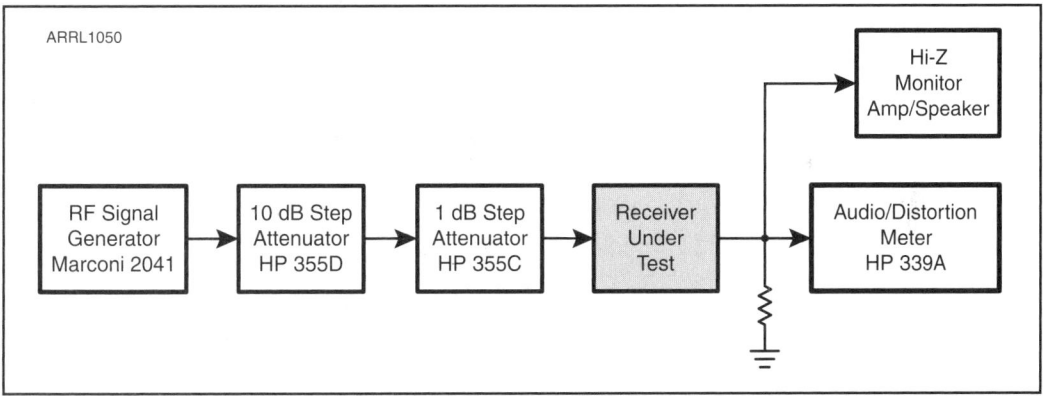

Figure 8.2 — AM and FM sensitivity test setup.

In the AM sensitivity test, the measurement is made using a distortion meter, and the 10 dB S+N/N AM signal level creates a distortion level of 31.8%. The signal generator output is increased until the receiver output level is equal to 31.8% on the distortion meter. The generator's output level, in µV, is noted as the AM sensitivity.

8.3 FM Sensitivity

FM sensitivity is measured as an FM signal level that is roughly 12 dB above the background noise. This FM signal also has internal receiver noise and distortion riding along with it, so the measurement is considered to be 12 dB signal plus noise plus distortion divided by noise plus distortion (you'll see this written as 12 dB SINAD).

The FM sensitivity measurement is made with the same test setup as before, shown in Figure 8.2. This time the receiver under test is adjusted for a 15 kHz filter bandwidth, the signal generator is switched to FM, and the deviation set to 3 kHz with a 1000 Hz tone.

In the FM sensitivity test, the signal generator output level is adjusted until a 25% distortion level is shown on the distortion meter, equaling a 12 dB SINAD level. The generator's output level, in µV, is noted as the FM sensitivity.

8.4 FM Adjacent Channel Selectivity

At what signal level will a single adjacent signal degrade your FM reception? Finding the answer to that question is the purpose of the FM adjacent channel selectivity test. **Figure 8.3** shows an FM adjacent channel selectivity test setup. This test is performed using two signal generators that are combined to simulate the desired signal at the 12 dB SINAD level and an "offending signal," 20 kHz away.

Each generator is set for FM mode and 3 kHz deviation. The modulation tone of Generator 1 is set to 1000 Hz, and Generator 2 is set for 400 Hz. Generator 1, the desired signal, is adjusted for the 12 dB SINAD level. Using a distortion meter, Generator 2's signal level is increased until the desired signal degrades to a 6 dB SINAD level (50% distortion). The level of Generator 2 is noted and the difference between the 12 dB SINAD level and the level of Generator 2 determines the adjacent channel selectivity. The

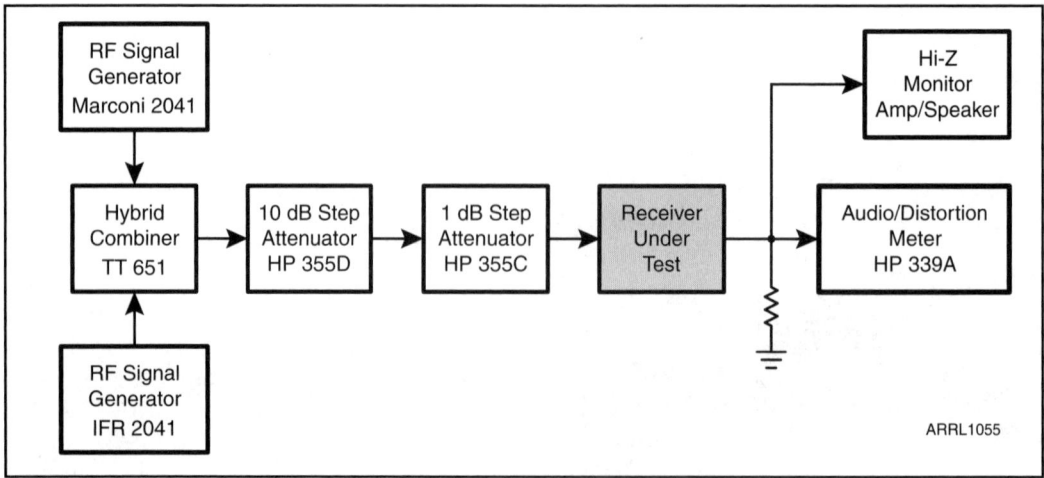

Figure 8.3 — FM adjacent channel and two-tone, third-order dynamic range test setup.

greater the FM adjacent channel selectivity figure, the better. Above 70 dB is good, 50 dB or less is poor.

While the serious contester or DXer may dismiss FM operation, there are a handful of transceivers that cover the HF, 6 meter, 2 meter, and 70 cm ham bands. To the casual operator who wants an all-in-one transceiver, FM performance is important. One positive aspect of operating FM is the mode's immunity to noise from manmade sources, such as power lines, although the overall sensitivity is reduced due to the raised noise floor.

8.5 FM Two-Tone Third-Order Intermodulation Distortion Dynamic Range

The undesired IMD signal created by two close, equally spaced, strong adjacent signals was explained in Chapter 6, but we will revisit here how it relates to FM operation. The purpose of the FM two-tone third-order IMD dynamic range (3IMD DR) measurement is to determine the possibility of interference to the desired signal created by two strong adjacent FM signals. It's possible that a radio amateur may be in the vicinity of two strong adjacent repeaters, say 15 and 30 kHz away from the received frequency. When both repeaters are transmitting, a false, garbled signal is received. An undesired signal can be generated within the receiver from stations that are 20 and 40 kHz away, or 10 and 20 MHz away, or any other equally spaced combinations.

The FM 3IMD DR measurement is made with two combined signal generators, using the same test setup as used for the FM adjacent channel selectivity test (Figure 8.3). Each signal generator is set at a fixed level, set to the FM mode, with a step attenuator controlling the combined output. The level of this combined output is adjusted to create a false IMD signal at a level of 12 dB SINAD. That level is compared to the FM sensitivity figure in dBm (see

Some Voice Mode Tips

I've talked with many new hams whose only requirement from their first HF transceiver was to get them on the air and talking, using SSB. I'm always happy to tell them about other voice modes, because both AM and FM can be fun. There are times when AM sounds like you are using a high-quality intercom. It is easy on the ears, with a quality that rivals some AM broadcast stations. Ten meter FM has greater local reach than 2 meters, and during periods of high solar flux levels, it can be ripe with DX from 29.2 to 29.7 MHz.

New Technician class licensees, who are limited to 28.3 to 28.5 MHz, SSB only, can enjoy the AM and FM modes on 6 meters and up. I find modern transceivers very useful for monitoring 50.4 MHz, the 6 meter AM calling frequency. I can catch my local friends and chat while I wait for one of my old 6 meter AM tube transceivers to warm up and stabilize. As with SSB on 6 meters during times of sporadic-E skip, 50.4 MHz can be a hotbed of activity. Wise VHF contesters stop by 50.4 MHz to make a few extra contacts to push their total scores up. (I just gave away one of my secrets!)

Table 8.1), and the difference is determined to be the FM two-tone 3IMD DR.

The measurement is taken with the generators set at 20 and 40 kHz above and 20 and 40 kHz below the desired frequency for this test. The results of both measurements should be about the same, but as always, the worst of the two figures is reported. The test is repeated at 10 MHz spacing to check for susceptibility to interference from signals from other services outside the ham bands (public safety, pagers and so on). An FM 3IMD DR of 70 dB is considered good, 80 dB and above is excellent, 50 dB is mediocre, and below that is poor.

When I was a young lad I hooked my new, fully synthesized handheld 2 meter transceiver to a new, highly elevated base station vertical antenna, thinking I could communicate over greater distances. I lived on a high hill, less than 10 miles from the RF hot spot of Hartford, Connecticut. On many popular repeater frequencies, all I could hear was a garble of pagers, commercial dispatch stations, and the police. I suspect the FM 3IMD DR at the 10 MHz spacing of that handheld, designed for a short flexible antenna, was poor.

Chapter 9

Filters

In previous chapters we've mentioned bandwidth or filter settings for various tests. Here's more information about these critical components that help the receiver hear one signal at a time and keep nearby signals from interfering.

Manufacturer's Specifications
IF/audio response: Not specified.

Measured in the ARRL Lab
Range at –6 dB points, (bandwidth)[‡]:
  CW (500 Hz): 450-947 Hz (497 Hz)
  Equivalent Rectangular BW: 501 Hz
  USB (2.4 kHz): 164-2306 Hz (2142 Hz)
  LSB (2.4 kHz): 157-2295 Hz (2138 Hz)
  AM (6 kHz): 79-2696 Hz (5234 Hz).

Notch filter depth: Not specified.

Manual: >70 dB; auto: >70 dB,
  attack time: 100 ms.

[‡]Default values; bandwidth and cutoff frequencies are adjustable via DSP. CW bandwidth varies with PBT and Pitch control settings.

9.1 Filter Evolution

Since the beginning of radio communications, there has always been the need for single signal reception. One hundred years ago, receiver selectivity was not very good, and old-time operators experimented with ways to narrow the range of frequencies heard through headphones. Early on, this often took the form of cutting out shapes, or slots, in each headphone's metal diaphragm. Fortunately, as receiver technology evolved, sets such as the one shown in **Figure 9.1** incorporated major improvements to the older designs.

First, operators began placing different fixed condensers (capacitors) in line with the audio, configured to be a variable low-pass audio filter (in this case, switchable via the TELEPHONE CONDENSER control shown in

Figure 9.1 — The author's IP-501 marine receiver, which was "state of the art" in 1920.

Figure 9.2). This feature limited the upper audio frequencies to reduce interference, at least somewhat.

IF crystal filters introduced in the late 1930s were still another major improvement, making single signal reception in the CW mode easier. This type of filter always adds to the price of a receiver (or transceiver), because the crystals must be high quality and considerable time and effort is goes into matching them in a lattice configuration for best performance. As technology improved, receivers incorporated switchable IF filters at bandwidths appropriate for different modes. For example, a receiver might incorporate a 500 Hz crystal filter for CW and RTTY, a 2.4 kHz filter for SSB and a 6 kHz filter for AM. Later designs included provision for several filters for each mode, as well as switchable filters at each IF.

Some transceivers use Collins mechanical filters instead of crystal filters to narrow the received bandwidth. In either case, the full complement of optional filters added significantly to the cost of a typical transceiver. For more information on filters, see "Transceiver Features That Help You Beat Interference" by Dave Newkirk, WJ1Z, in the March 1991 issue of *QST*.

The next major filter improvement was made in the 1990s with the introduction of digital signal processing (DSP) technology. DSP is used today to provide highly

Figure 9.2 — A close up of the IP-501 shows an early attempt of filtering out high-pitched audio tones with the control labeled, "Telephone Condenser."

adjustable and highly effective bandwidth filtering without the need to add expensive crystal filters for each bandwidth selection. All recent HF transceivers include extensive DSP filtering options.

The latest major technical advance was the incorporation of selectable roofing filters in mid- to high-end performance transceivers. At first, typical transceivers included a single 15 or 20 kHz wide crystal filter in the first IF and relied on DSP filtering at the second IF for narrower bandwidths. Roofing filters — narrow bandwidth crystal filters at the first IF — offer significant performance improvements as discussed in the next section.

Radio amateurs in the market for an HF transceiver should carefully consider their operating needs regarding filtering. The serious operator will want as many filter options as possible, in order to attempt single signal reception during crowded band conditions.

9.2 Roofing Filters

As your home's roof is the first line of defense against rain, a roofing filter is your transceiver's first line of defense against one or more strong adjacent signals. The roofing filter is located after the first mixer and is there to narrow the passband of the first IF. This helps reduce overload in stages further down the receiver chain where additional bandwidth filtering occurs (typically through DSP in current designs). So we see that a receiver's bandwidth is not determined entirely by the roofing filter, but the roofing filter is a key part of the overall system. To continue our analogy, while a roofing filter can help receiver performance considerably, if it rains hard enough, the roof will still leak.

A 3 kHz wide roofing filter will improve the two-tone, third-order IMD dynamic range (3IMD DR) and blocking gain compression dynamic range (BGC DR) at 20 kHz and 5 kHz spacing. This improvement occurs because all but the strongest adjacent signals 3 kHz or more away have been reduced before the adjacent signal arrives at the first IF amplifier. A 3 kHz roofing filter is most valuable if the operator is using the SSB mode and has a large antenna array that provides strong signal voltages from adjacent stations at the antenna jack.

For the serious CW or digital operator, a roofing filter of 1000 Hz bandwidth or less can improve 3IMD DR at close spacing. Performance improvements are evident in the Lab's dynamic range testing at 2 kHz spacing. In the best receivers with good quality narrow roofing filters, test results are very close at all three spacings.

For the casual operator with a simple dipole or vertical antenna, a narrow roofing filter is usually not needed. Signals levels from nearby stations typically are not strong enough to degrade receiver performance. Of course that's not the case if the operator lives near another amateur station that is active on the same ham band.

Some mid- to high-performance receivers offer a 15 kHz roofing filter as an option. This is useful while operating in the AM or FM modes, where a narrower roofing filter will degrade the audio and a wider IF passband is needed.

Aftermarket kits are available to replace the 15 kHz or 20 kHz first IF filter in older transceivers with a narrow roofing filter of 3 kHz or so bandwidth. Keep in mind that while enhancing performance of a receiver while listening to SSB and CW signals on a crowded band, the narrow roofing filter will make listening to AM and shortwave broadcast stations unpleasant.

9.3 Digital Signal Processing Filters

DSP filtering takes place after the last IF stage (the second IF in current designs), or in the audio stage where analog signals are converted to digital information for manipulation, and then back to analog for the user to hear. The passband can be expanded or narrowed or shifter higher or lower at the user's discretion. In some cases bandwidth is adjustable in 50 Hz or 100 Hz steps over a very wide range, and filter shape is adjustable as well. Some modern transceivers include graphic displays of the currently selected filter characteristics (see **Figure 9.3**).

On mobile and portable transceivers with DSP filtering, adjustments

Figure 9.3 — The graphic display on Icom's IC-7600 shows quite a bit of information about filter settings in the lower part of this screen.

are made via a menu function. On HF transceivers with more front panel space, there will be DSP filter control knobs or switches to adjust bandwidth.

Many software defined receivers (SDR) use DSP technology exclusively to adjust receiver bandwidths and to provide a visual graphic display on the front panel or an external monitor.

9.4 IF Crystal Filters

Prior to the current generation of transceivers using roofing filters at the first IF in conjunction with DSP filters at the second IF, typically a new transceiver would arrive from the manufacturer or distributor with a stock 2.4 or 2.7 kHz IF filter and provision for one or more optional filters with different bandwidths. In addition to IF filtering, most previous generation transceivers included some audio DSP filtering, and the combination of both types of filtering provides enough reduction of interference for the casual radio amateur. Although newly designed transceivers no longer use this technology, a number of models with optional filters are still available new, and many more are available on the used market.

Additional optional IF filters can be rather expensive ($100 or more each), but they can improve the selectivity on the desired mode. Some manufacturers or distributors will pre-install an optional filter for the purchaser, or the purchaser may have to install the optional filter by removing one or both covers and plugging in the filter or soldering it in place. The most common optional filters are 1.8 kHz for SSB and 500 Hz for CW reception. Additional choices may include 250 Hz for CW and 6 kHz for AM.

There is a tradeoff when using narrow bandwidth crystal filters. For instance, some 500 Hz or narrower CW filters can have a ringing quality and may not provide a pure audio tone. A 1.8 kHz SSB filter will roll off the high-frequency audio components, making the speaker audio sound muddy. These are compromises to consider in your quest for single signal reception during crowded band conditions. The *QST* Product Review author will usually comment any issues with the audio quality of installed filter choices.

9.5 Filter Measurements

Let's get back to the numbers presented in the Product Review data table. Toward the bottom of the receiver section is series of measurements under the heading "IF/audio response." Usually these numbers are not specified by the manufacturer, but the ARRL Lab performs measurements for the CW, SSB and AM modes.

The purpose of this measurement is to verify that the receiver's audio frequency response matches the selected filter bandwidth. IF/audio response is measured with the DSP filtering at the default settings, with a 500 Hz bandwidth for CW, 2.4 kHz for SSB and 6 kHz for AM. If the receiver includes selectable roofing filters, a mode-appropriate filter is selected (in the example transceiver, 600 Hz for CW and 3 kHz for SSB). Upper sideband (USB) and lower sideband (LSB) are measured and reported separately.

The test setup is shown in **Figure 9.4**. For the SSB and CW measurements, a signal generator is used without modulation, with the generator's frequency adjusted to provide a relative audio reference level on an audio meter in the middle of the audio passband. The generator frequency is adjusted higher and lower to determine at what frequencies the audio at the edges of the passband decrease by 6 dB (the "6 dB down" points). The low and high frequency points of the filter bandwidth are measured by feeding the speaker output to a frequency counter.

AM response is measured by simply changing the modulated tone of the signal generator from lower to higher to find the point at each edge of the audio passband where the signal is reduced by 6 dB. The difference between the highest and lowest frequency is then multiplied by two to allow for both audio sidebands.

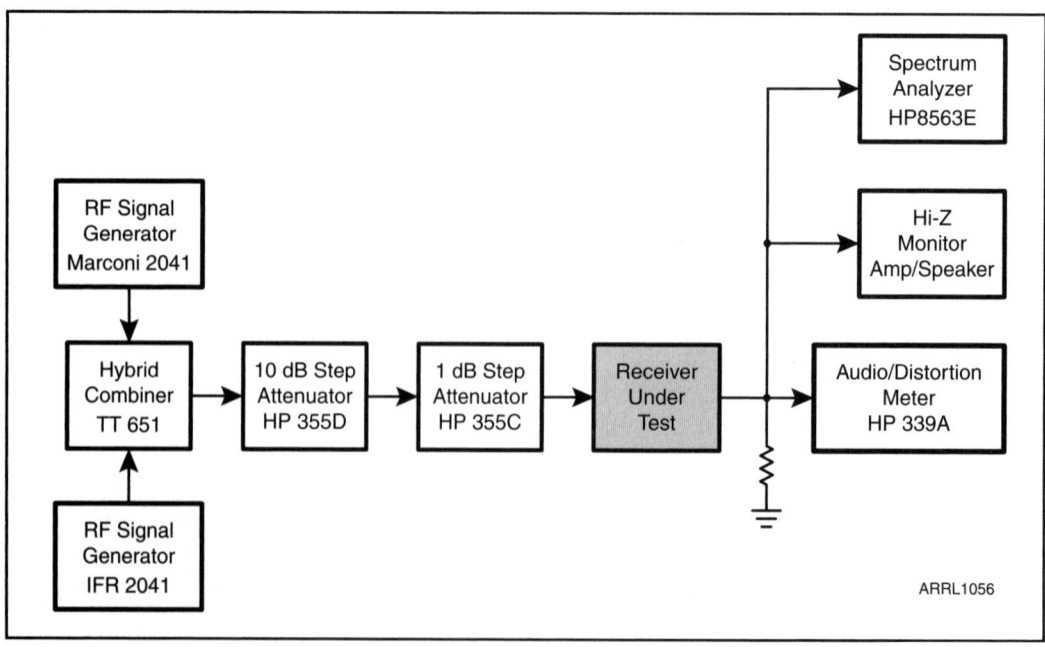

Figure 9.4 — Filter response test block diagram.

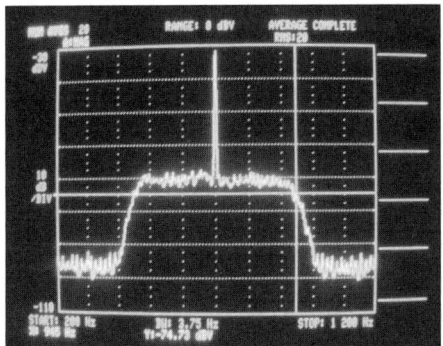

Figure 9.5 — Filter response as seen on a signal analyzer. The receiver is tuned to a 1 µV signal, with an audio signal entered at 700 Hz.

After the CW bandwidth we report Equivalent Rectangular Bandwidth (ERBW). This is the width that the filter would have if it passed the same noise power and possessed the "ideal" shape of vertical sides and a flat passband response. In this case, with the 500 Hz filter setting, actual bandwidth of 497 Hz and ERBW of 501 Hz, we are right on target. In some transceivers, the differences are much more pronounced.

A filter response that is ±10% of the selected filter bandwidth is acceptable (**Figure 9.5**). Keep in mind that these are the 6 dB down points. At 6 dB down, the human ear perceives a drop in audio level to one quarter of the highest audio level in the middle of the audio passband. Years ago, we started doing this test because some transceivers had severely restricted audio passbands on SSB, reducing intelligibility. Even though a 2.4 kHz crystal filter was selected, additional audio filtering (high-end roll-off to reduce hiss, for example) might reduce the effective bandwidth to, say, 1.5 kHz. In general, today's receivers are much better

There are also "skirts" on either side, with the most observable being on the higher frequency side where audio will still bleed through to some extent. Shallow filter skirts (a more gradual drop in attenuation) will result in more interference. High-performing receivers will have a selection of soft, medium and sharp filter skirts.

In my testing, I pass any reports of unusual filter characteristics along to the reviewer, who will include my comments in the review article. I've also observed very sharp drop-offs on the audio passband with SDR receivers, which is very desirable.

9.6 Notch Filters

A notch filter attenuates signals over a very narrow range, with maximum attenuation at a single frequency. Typically, notch filters are used to reduce or eliminate one or more audio tones (carriers) while operating in the SSB mode, and the frequency of the notch can be adjusted manually. In modern transceivers, DSP-based automatic notch filters can seek out and attenuate one or more carriers in the passband at the touch of a button.

An interfering audio tone usually comes from someone who is transmitting while adjusting an amplifier or an antenna tuner. On 40 meter phone, it might be the carrier of a shortwave broadcast station operating above 7.2 MHz. That tone is annoying and sometimes even painful, but

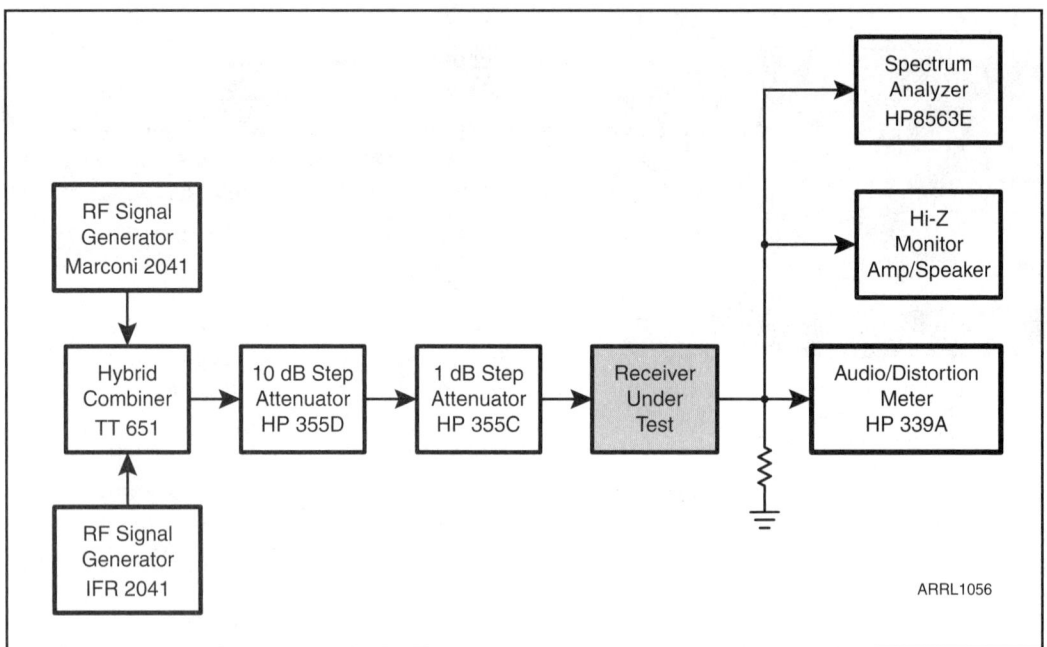

Figure 9.6 — Notch filter depth test block diagram.

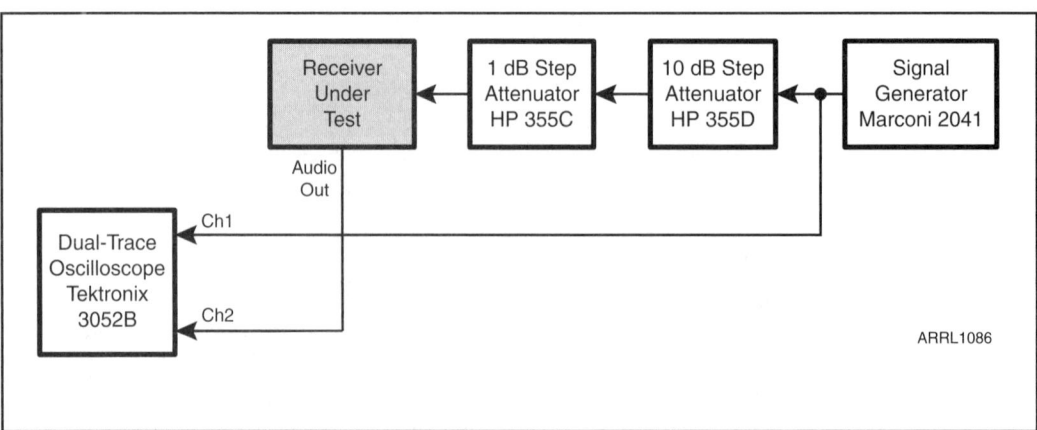

Figure 9.7 — Test setup for measuring the attack time of an automatic notch filter.

a manual or automatic notch filter can reduce (null) the level of the tone significantly with little or no impact on the desired audio signal, allowing you to carry on with your contact.

For transceivers with notch filters, the Lab tests notch depth (maximum attenuation of an interfering tone) in both the manual and automatic modes (if so equipped). The test setup is shown in **Figure 9.6**. Two 20 meter input signals, spaced 1200 Hz apart, are applied to the receiver under test and adjusted for equal output amplitudes as observed on the signal analyzer. The receiver notch filter is then set to null the undesired upper audio tone. Both tones are observed on the spectrum analyzer display. The depth of the notch filter is then determined by comparing the difference in levels between the two tones. Of course the greater the notch depth, the better the elimination of an interfering signal.

In addition, we measure and report automatic notch filter "attack time" (the time it takes for the tone to go away) for single or multiple audio tones, if applicable. The test setup for this is shown in **Figure 9.7**. In this test a signal is fed into the receiver and also one channel of a dual-trace oscilloscope, where it is used to trigger the sweep. Receiver output is monitored with the second oscilloscope channel. The oscilloscope is placed in the single sweep storage mode and the signal generator keyed. The notch filter attack time is the time delay from the point the signal appears until it is notched by at least 50% and is based on the average of several measurements.

By the way, I've heard from many amateurs who called to say they were having trouble copying CW signals. The first question I ask these days is, "is the notch filter button on?" After some fussing on the other end, I'll hear, "beep" and, "That's better." Remember to turn off the notch filter while listening to CW stations.

Chapter 10

Receiver Audio Output

A receiver must be comfortable to listen to for extended periods. This test measures the amount and quality of a receiver's audio output.

Manufacturer's Specifications
Receiver audio output: 2.5 W into 4 Ω at 10% THD.

Measured in the ARRL Lab
2.6 W at 3.2% THD into 4 Ω (maximum audio). THD at 1 V RMS: 0.4%.

10.1 Maximum Audio Output

This test checks to see if the receiver meets the manufacturer's specifications for maximum audio output and total harmonic distortion (THD) into a specified (speaker) load. If unspecified, the power is measured with an 8 Ω load at 10% THD. The power output into a speaker can range from a few hundred milliwatts for simple kit transceivers to several watts for

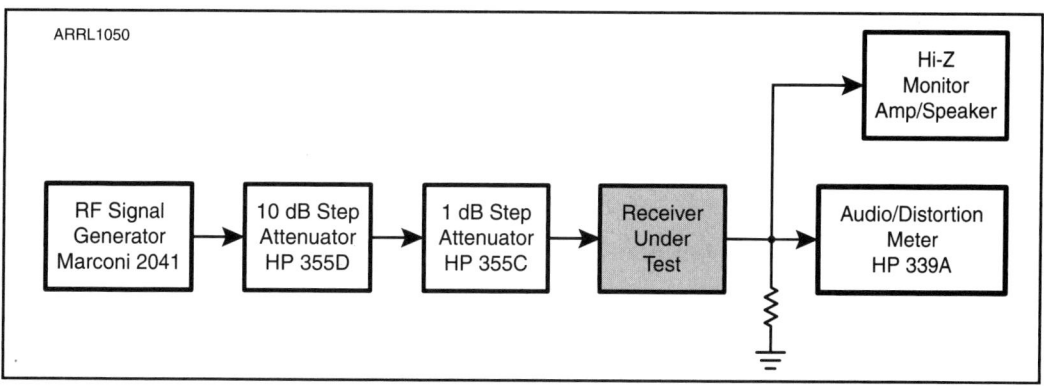

Figure 10.1 — Audio distortion and the maximum audio power output test setup.

a mobile transceiver. A typical HF and 6 meter transceiver has at least 2 W of maximum audio power output. Note that the headphone jack is not used because it may not develop the full power available from the speaker jack.

Figure 10.1 is a block diagram of the audio distortion and maximum audio power output test setup. A signal is fed into the receiver and the frequency adjusted until the audio tone is approximately 1000 Hz. The receiver's audio gain is increased until the audio/distortion meter indicates the manufacturer's specified THD. If unspecified, the audio gain is increased until the THD reaches 10%. If the THD is less than 10% at maximum volume, as it is in the example receiver, that's noted in the test data.

Amateurs who want to use their transceivers in a vehicle or in a portable environment need to have enough audio power output to overcome road and wind and other noise. For this application, the more audio output power, the better.

10.2 Tweaking a Classic Test

Who really listens to their speaker audio cranked up all the way to a point of 10% THD? I wondered this a few years back and decided it would useful to measure a receiver's audio distortion at a power level that most radio amateurs actually use in a home station environment. After some subjective testing and consultation with other ARRL staff members, I decided to add a second measurement — audio distortion at the 1 V level (1 V RMS of audio across the speaker). This represents a reasonably comfortable home station listening level and is indicative of audio quality, rather than quantity.

This measurement is important for anyone who operates for hours on end or who enjoys ARRL Field Day or contesting. Even a few percent of audio distortion can be tiresome, adding to fatigue or making accurate copy difficult.

The radio amateur looking to purchase an HF transceiver should keep an eye out for a low distortion figure at a speaker volume of 1 V RMS. At 1 V RMS, a THD of 0.5% or lower is good; THD of 1% to 2% is mediocre.

Chapter 11

Noise Reduction

Noise from power lines, motors, automobile ignitions and other manmade sources can range from annoying to debilitating. Several receiver features can help to reduce interference from external noise sources.

Manufacturer's Specifications	Measured in the ARRL Lab
DSP noise reduction: Not specified.	Variable, 30 dB maximum.

11.1 Automatic Noise Limiter

Noise has always been troublesome for the radio amateur. Old-timers used to tell me the horror stories of early electric motors, automobile ignition systems, and sparking trolley poles dragging across long, overhead electrified wires. In the 1930s, some receivers used crude noise limiters that clipped the peaks of the audio pulses and reduced noise somewhat, but usually at the expense of audio quality. Later on, automatic noise limiters (ANLs) became available as a feature in many communications receivers. ANLs helped, but they did not eliminate manmade noise sources entirely.

11.2 Noise Blanking

Noise blanking is useful for reducing or eliminating manmade repetitive pulse noise, such as ignition noise and, to varying degrees, power line noise. A noise blanker, however, may not be effective in reducing static interference caused by lightning. A noise blanker control may be a simple on/off switch with a fixed blanking level, or it may feature an adjustable blanking level knob.

Noise blanking circuitry works best on short repetitive pulses. Adding width to the pulses or other shapes will lower the effectiveness. Engaging the noise blanker, however, can add to unwanted effects, such as audio distortion products. A noise blanker, when turned on, can also make strong CW signals appear to have key-clicks up and down an amateur band, for example.

Mobile operators and radio amateurs living in congested areas with multiple manmade noise sources should carefully read the Product Review author's comments on the effectiveness of the noise blanking feature of a given transceiver. We are in the process of developing noise blanker tests.

Noise blanking can be effective on other kinds of signals. In the mid 1970s, a very strong, manmade noise source appeared up and down the HF spectrum and frequently invaded a substantial portion of an amateur band. It became known as the Woodpecker and it sounded like woodpecker hammering away on a metal rooftop. The Woodpecker turned out to be a Russian over-the-horizon radar that operated in the HF spectrum and used the ionosphere to obtain a greater range than typical line-of-sight radar. When transmitting, the Woodpecker obliterated radio reception on or near frequencies in the ham bands. Manufacturers soon responded with new noise blanking circuits that helped to reduce the effects of the Woodpecker. Fortunately, the Russian Woodpecker has been silent since the end of the Cold War, though similar signals from other countries are still heard once in a while.

11.3 DSP Noise Reduction

During the late 1990s, DSP noise reduction was added to mid- to high-end HF transceivers. DSP noise reduction is a major design enhancement that greatly reduces manmade noise and natural interference. It uses the DSP circuitry to process signals with mathematical algorithms. The system recognizes what is noise and what is desired audio, and the desired audio is passed along to a digital-to-analog converter (DAC) to enable the sound to come out the speaker.

With DSP noise reduction there is, however, a bit of a tradeoff between an improved signal-to-noise ratio and the quality of the speaker audio. As DSP speed and processing power have improved, so has the audio quality of a signal with noise reduction applied. It's now a standard feature on desktop HF transceivers. Thanks to the miniaturization of electronics, some small mobile and portable HF transceivers have DSP noise reduction available, usually through a menu function.

Figure 11.1 is a block diagram of a DSP noise reduction test setup for HF transceivers with DSP noise reduction capability. An uninterrupted, unmodulated CW signal from a signal generator is mixed with a wideband noise source and fed to the transceiver's antenna jack. Using a signal analyzer, the difference between the CW (reference) signal and the noise

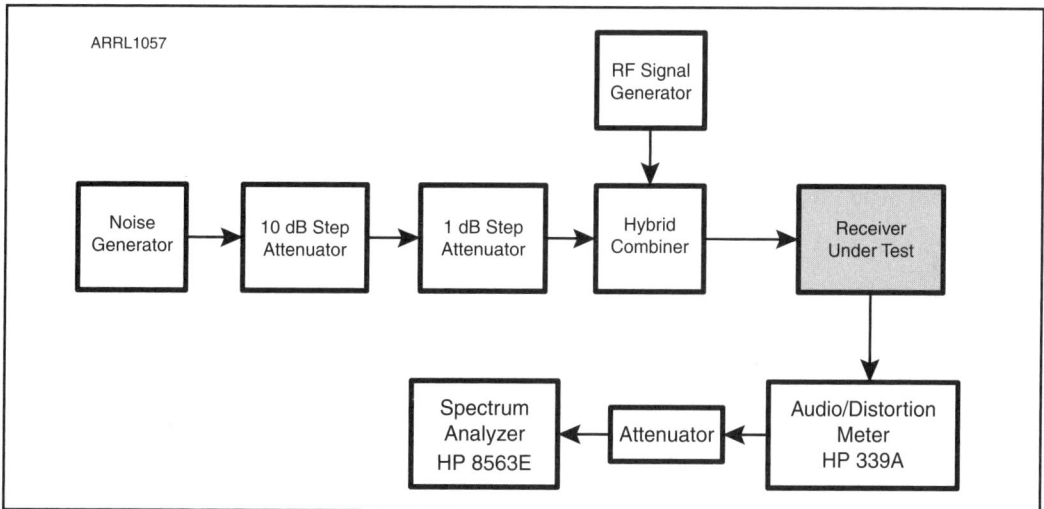

Figure 11.1 — DSP noise reduction test block diagram.

level is measured with the noise source on and then measured again with the noise source off.

Typical levels of DSP noise reduction, as measured at the ARRL Lab, are 10 to 20 dB. At up to 30 dB, the example receiver's noise reduction is quite good.

For transceivers without built-in DSP noise reduction, aftermarket external speakers with DSP noise reduction are available. When adjusted to the maximum DSP noise reduction setting, voice transmissions tend to have the "space alien" sound observed with some DSP circuitry — voice characteristics are significantly altered, but are at least audible in high-noise environments.

Choosing an HF transceiver with DSP technology has many advantages, including the improved noise reduction explained here.

Chapter 12

Other Receiver Measurements

This chapter covers the final few receiver measurements included in a typical HF transceiver data table.

Manufacturer's Specifications
S meter sensitivity: Not specified.

Spectral display sensitivity: Not specified.

Squelch sensitivity: Not specified.

Measured in the ARRL Lab
S9 signal at 14.2 MHz, preamp off/1/2: 94.3/24.8/9.2 µV.

Preamp off/1/2: –100/–113/–120 dBm.

At threshold: SSB (preamp off), 9.22 µV; FM, 29 MHz (preamp 2), 0.42 µV; 52 MHz (preamp 2), 0.33 µV.

12.1 S Meter Sensitivity

The S meter is a point of reference to help quantify the strength of an incoming signal. The industry standard is that S9 equals a 50 µV input signal (50 µV is equal to –73 dBm; refer to Table 8.1 in Chapter 8 for conversion information). The ARRL Lab reports the level, in microvolts, at which S9 is reached on transceivers that are equipped with an S meter. A 20 dB over S9 signal, for instance, is at the –53 dBm level (–73 dBm plus 20 dB). On the lower end of the scale, each S unit is roughly equivalent to 6 dB. An S8 signal level is at –79 dBm, for example (6 dB weaker than –73 dBm).

The purpose of the S meter sensitivity test is to measure how well the S meter feature in the receiver under test conforms to the S9 = 50 µV standard. The measurement of S meter sensitivity is performed by simply placing a signal generator at the antenna jack and adjusting the generator

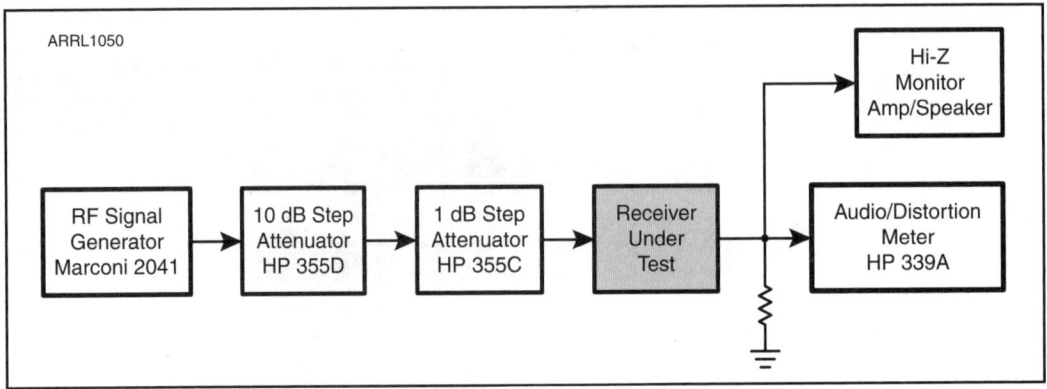

Figure 12.1 — Test setup for S meter sensitivity, spectral sensitivity and squelch sensitivity.

level until the S meter of the receiver under test, with all preamps off, reads exactly S9. The signal generator output level, in μV, is noted. The test is repeated with each preamp setting. **Figure 12.1** shows the test setup, the same one used for receiver sensitivity and other tests described in previous chapters.

Many transceivers will read S9 with a signal input less than 50 μV (–73 dBm) with the preamp off, then will read considerably higher with a preamp turned on. That's certainly the case with the example transceiver. The signal level required for an S9 meter reading varies greatly with preamp settings. This doesn't mean the on-air signal received at the antenna jack magically gets higher! A few manufacturers have addressed this behavior. In their transceivers, the S meter reading is the same regardless of preamp setting.

On the air, I always give signal reports with the preamp turned off. That gives the operator at the other end an idea of the signal voltage being delivered to my antenna jack.

Manufacturers of amateur transceivers are gradually phasing out the traditional analog S meter (**Figure 12.2**) in favor of a graphic display (**Figure 12.3**). On smaller transceivers, the graphic S meter is a segmented part of an LCD display that at times can be difficult to read at arm's length (**Figure 12.4**). For displays with more real estate, a graphic recreation of a traditional S meter is

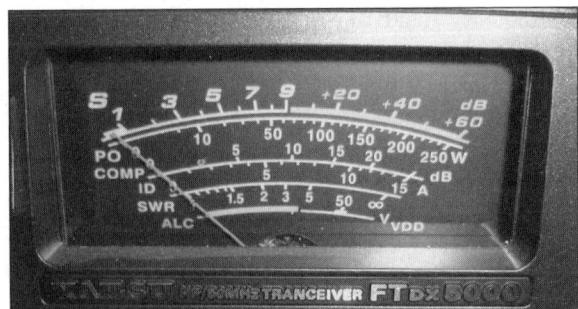

Figure 12.2 — In this Yaesu FTDX1200 transceiver, the analog S meter is also used on transmit to measure compression, ALC, power output, drain current, drain voltage and SWR.

Figure 12.3 — As graphic displays on higher end transceivers such as the Kenwood TS-990S become more sophisticated, manufacturers are able to replicate an a traditional analog S meter in graphic form. As with the analog meter shown in Figure 12.2, this display also monitors a number of parameters on transmit.

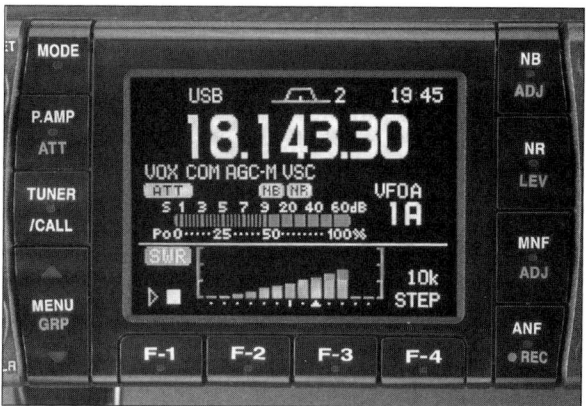

Figure 12.4 — Smaller transceivers, such as the Icom IC-7000, incorporate a bar graph S meter in the display.

featured. Some SDR transceivers include both an S meter display and a dBm level indicator.

The transceiver's S meter is generally used for transmitting purposes as well. You'll find automatic level control (ALC), RF power output and standing wave ratio (SWR) among the typical metering functions.

Keep in mind that after you purchase your HF transceiver, you'll be spending a lot of time staring at your S meter. Make sure you're happy with its readability and that it will not cause fatigue. The Product Review author will comment on any issues with metering.

12.2 Spectral Sensitivity

Spectral displays (also known as panadapters or spectrum scopes) offer a visual display of signals over a range of frequencies. See **Figure 12.5**. In some transceivers, the spectral display will show moderate to strong signals well, but will not show weaker signals that are clearly heard from the speaker. Observing this for the first time, I created a new test called "Spectral Sensitivity." The purpose of the test is to measure and report the weakest signal that "pops up from the grass," with the grass being receiver noise along the bottom of the display.

The test setup the same as that shown in Figure 12.1. For this test, the signal generator is adjusted to a level at which a signal is clearly visible above the grass, about 3 to 5 dB above the tops of the grass blades (to continue the

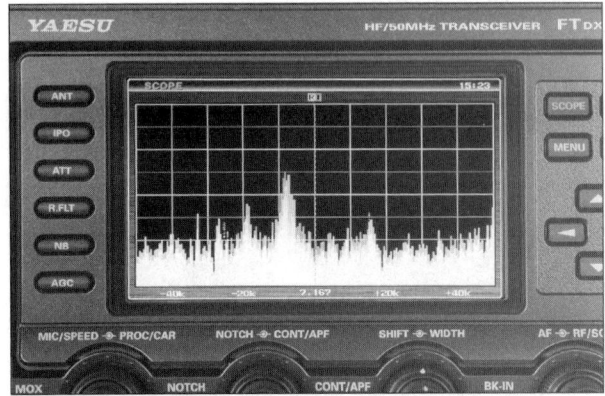

Figure 12.5 — The spectral display included on many current transceivers shows signals present in a selected portion of a band (in this case, 50 kHz above and below the operating frequency). If spectral sensitivity is low, weaker signals will not be displayed.

analogy) while observing a 100 kHz section of a ham band, if applicable. This measurement is subjective since another person may see the signal in the grass better than I can, but the difference would only be slight.

If you are considering the purchase of a receiver/transceiver with a spectral display, compare the MDS figures with the spectral sensitivity figures to see if they are close. A difference of 6 dB or less is good. In the example transceiver, the difference is as high as 27 dB with the preamp off. All signals at an S4 level (–100 dBm) or weaker will be missing.

Another option to visually see signal strengths is a "waterfall." A waterfall display is commonly used as a tuning aid in digital modes such as PSK 31. A very faint "trace" shown on a waterfall can be made from a received signal at or below the MDS level. As the received signal level

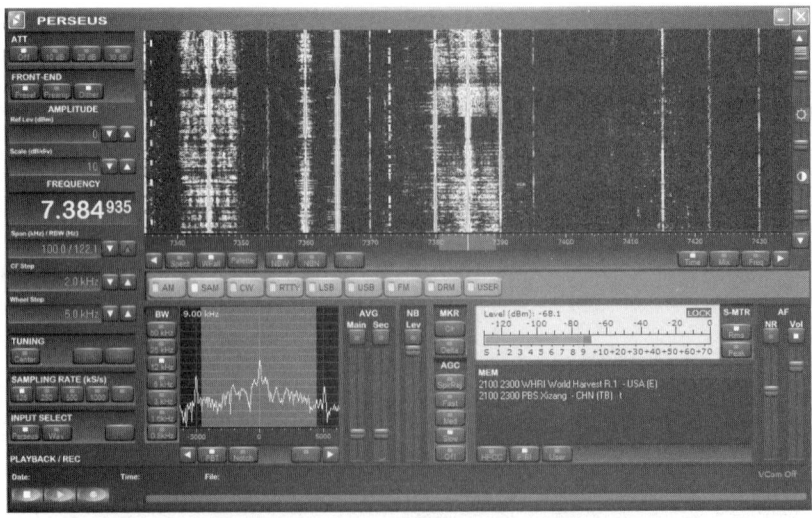

Figure 12.6 — Software defined radios such as the Perseus SDR receiver often include a waterfall display option as part of the companion software. In addition to the waterfall display, there is a close-in spectral display of the received signal in the lower part of the screen.

increases, the color of the trace changes. SDR transceivers/receivers typically offer both panadapter and waterfall displays of signals (**Figure 12.6**).

12.3 Squelch Sensitivity

A squelch control is included if a transceiver includes the FM mode or has an option for FM operation. Squelch sensitivity is measured in the FM and SSB modes to determine the level of the input signal required to break the squelch at the threshold point. Measured in microvolts, it is the input signal level at which the squelch opens and is held open (the "squelch threshold").

The purpose of the FM squelch sensitivity measurement is to determine if the squelch threshold is close to the FM 12 dB SINAD level. That's a level at which, by ear, the receiver audio has a fair amount of background noise, but the received signal is still strong enough to convey speech.

Squelch sensitivity is measured in the FM portions of the 10 meter and 6 meter bands (and higher frequency bands if applicable). The test setup was shown previously in Figure 12.1. For squelch sensitivity, the signal generator is set to FM with 3 kHz deviation and a 1000 Hz modulated tone. The receiver audio is adjusted for a comfortable listening level and the generator output level adjusted until the signal is just heard in the receiver. The receiver's frequency is adjusted for minimum distortion. Then the signal generator's output is switched off and receiver's squelch control is adjusted to just past the threshold point, closing the receiver. The generator is switched back on and the signal level is increased until the squelch opens and is held open. The signal generator's output signal level is then noted.

If the squelch sensitivity is higher than the 12 dB SINAD level (more sensitive — a lower value in μV than the 12 dB SINAD level measured in Chapter 8), the squelch can be prone to false trips. Low squelch sensitivity (less sensitive — a higher value in μV than the 12 dB SINAD level) will mean that intelligible signals may be missed. The squelch sensitivity is good if it's equal to or slightly lower than the 12 dB SINAD level for the band tested.

The SSB squelch sensitivity is performed at 14.200 MHz with the receiver set to USB mode and the signal generator's FM setting turned off. The test procedure is the same as for FM, except this time the receiver frequency is adjusted for peak signal response (SSB) and the generator output level is increased until the squelch opens momentarily and closes again.

12.4 Automatic Gain Control

AGC is a vital function of any radio receiver. In the 1920s, before the use of automatic volume control (AVC) was common, a radio listener had to adjust the volume level up or down, depending on the strength of the signal received. This worked okay for listening to broadcast stations during the day, but at night, signal strengths from distant stations varied. It was more perilous for the radio amateur who was tuned to a weak station, with the volume circuits at maximum — with no AVC, a strong station could pop up out of nowhere and blast the eardrums! To relive this experience today, try turning off a receiver's AGC while leaving the RF gain all the way up and tune across the band.

Today, AGC serves as the "leveler," allowing for fewer volume adjustments and protecting the operator's hearing. A receiver usually has a FAST, MEDIUM and SLOW AGC setting. Most receivers also have an AGC OFF setting, since some operators prefer to "ride the gain." Usually, we set the AGC at FAST for CW and at MEDIUM or SLOW for voice modes for comfortable listening. In some transceivers, AGC characteristics for each setting are adjustable.

Most of us do not even think about the action of the AGC, but under conditions of quick, transient pulses, such as an electrical switch being turned on nearby or static discharge being generated, the AGC will momentarily cause the receiver to go quiet, even though the transient is inaudible to the user. I now test for this, using a pulse generator that creates the same effect as a brief static discharge. If I observe any unusual behavior, I pass this along to the reviewer. A radio amateur in the market for an HF transceiver should consider any comments on unusual AGC action.

Section 3

Transmitter Performance

Amateur receiver performance has improved dramatically over the years. The addition of DSP and roofing filters has made single signal reception a reality in CW operation.

Are there advancements on the transmitter side as well? While operators of modern transceivers enjoy a wealth of specialized receiver features, what about the impact of their transmissions on other radio amateurs? For instance, have CW keying and SSB audio improved? How clean are the emissions when transmitting digital modes? The tests described in the following chapters will help you to be a considerate operator by running a clean station.

Just what should the radio amateur new to HF operation be concerned about as he or she weighs the purchase of an HF transceiver? The contester or serious DXer also needs the transmitted audio or CW waveform or digital signal to be of high quality for intelligibility and to avoid interfering with nearby stations on a tightly packed band.

In the following chapters, I will address the transmitter measurements and give some advice on these topics.

Table S3.1
Transmitter Specifications from a Typical ARRL *QST* Product Review Data Table for an HF/6 Meter Transceiver

Manufacturer's Specifications

Transmitter

Power output: 5-100 W, (2-25 W AM).

Harmonic suppression: >60 dB (1.8-29.7 MHz), >65 dB (50-54 MHz).
SSB carrier suppression: At least 60 dB.
Undesired sideband suppression: At least 60 dB.
Third-order intermodulation distortion (IMD) products: −31 dB @ 14 MHz, 100 W PEP.

CW keyer speed range: Not specified.
CW keying characteristics: Not specified.
Transmit-receive turn-around time (PTT release to 50% audio output): Not specified.
Receive-transmit turn-around time (tx delay): Not specified.
Composite transmitted noise: Not specified.

Measured in the ARRL Lab

Transmitter Dynamic Testing

HF and 50 MHz: CW, SSB, RTTY, FM, as specified within specified supply voltage range. AM, 10-100 W (carrier).[††]
62 dBc (worst case 10 meters), 50-54 MHz, as specified. Meets FCC requirements.
>70 dB.
>70 dB.
HF, 100 W PEP, 3rd/5th/7th/9th order: −27/−40/−42/−52 dB (worst case, 10 m); >−31/>−40/>−45/>−52 (typical).
50 MHz, 100 W PEP: −32/−38/−45/−61 dB
4 to 59 WPM; iambic mode A or B.
See Figures S1 and S2.
S9 signal, AGC fast, 36 ms.

SSB, 34 ms; FM, 30 ms.

See Figure S3.

[††]Carrier level must be lowered to 25% of PEP for proper AM operation, for example 25 W carrier for 100 W PEP.

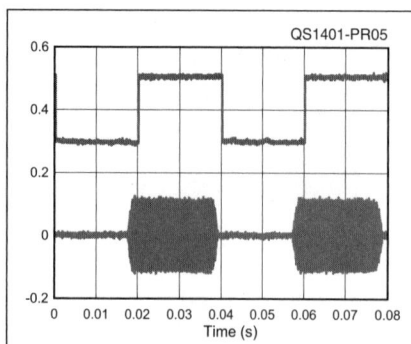

Figure S1 — CW keying waveform for the example receiver showing the first two dits in full break-in (QSK) mode using external keying. Equivalent keying speed is 60 WPM. The upper trace is the actual key closure; the lower trace is the RF envelope. (Note that the first key closure starts at the left edge of the figure.) Horizontal divisions are 10 ms. The transceiver was being operated at 100 W output on the 14 MHz band.

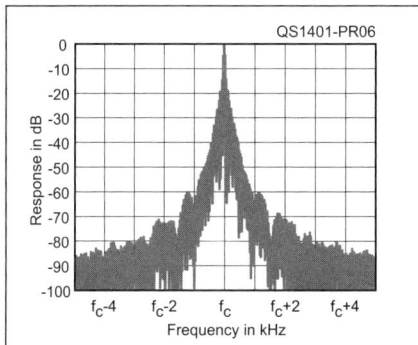

Figure S2 — Spectral display of the example transmitter during keying sideband testing. Equivalent keying speed is 60 WPM using external keying. Spectrum analyzer resolution bandwidth is 10 Hz, and the sweep time is 30 seconds. The transmitter was being operated at 100 W PEP output on the 14 MHz band, and this plot shows the transmitter output ±5 kHz from the carrier. The reference level is 0 dBc, and the vertical scale is in dB.

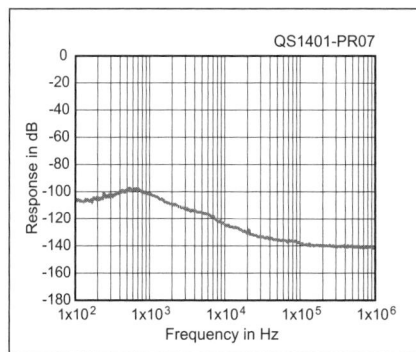

Figure S3 — Spectral display of the example transmitter output during composite-noise testing. Power output is 100 W on the 14 MHz band. The carrier, off the left edge of the plot, is not shown. This plot shows composite transmitted noise 100 Hz to 1 MHz from the carrier. The reference level is 0 dBc, and the vertical scale is in dB.

Chapter 13

Power Output, Spurious Signals and Harmonic Suppression

How much power does the transmitter put out on the various modes? Does it meet FCC requirements for spectral purity? We'll check those numbers in this chapter.

Manufacturer's Specifications
Power output: 5-100 W, (2-25 W AM).

Harmonic suppression: >60 dB (1.8-29.7 MHz), >65 dB (50-54 MHz).

Measured in the ARRL Lab
HF and 50 MHz: CW, SSB, RTTY, FM, as specified within specified supply voltage range. AM, 10-100 W (carrier).[††]
62 dBc (worst case 10 meters), 50-54 MHz, as specified. Meets FCC requirements.

[††]Carrier level must be lowered to 25% of PEP for proper AM operation, for example 25 W carrier for 100 W PEP.

13.1 RF Power Output

For all transceivers tested at the ARRL laboratory, the minimum and maximum RF power output is measured with an RF power meter of known accuracy (**Figure 13.1**) using the test setup shown in **Figure 13.2**. The RF power output tests are performed on each amateur band the transceiver covers, as well as on each mode of operation. For SSB, PEP output power is measured using a two-tone generator at the transmitter's microphone jack. No modulation is used in the AM and FM modes.

Tests are conducted using a 13.8 V dc power supply if the transceiver

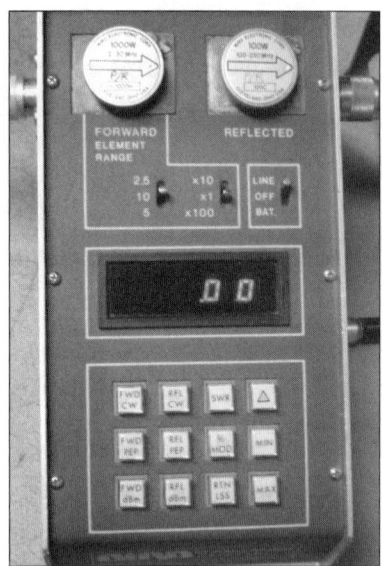

Figure 13.1 — A Bird 4391 power meter used for measuring RF power output.

has no internal 120 V ac supply. As mentioned in Chapter 2, most transceivers that require external power supplies are specified to operate over a range of voltages. The Lab also checks RF output at the minimum specified operating voltage, and if there is a drop in power output that's reported in the table as well.

Overall, today's modern transceivers meet the manufacturer's specified RF power output figure, typically 100 W or 200 W. Older transceivers that used vacuum tubes in the final amplifier frequently did not meet the specified RF power output on the higher HF bands such as 15 and 10 meters. A radio amateur looking to purchase an HF transceiver should check out the RF power output measurements in the Product Review data table to be sure that the transceiver makes the rated RF power output.

While it is advantageous to operate at the full specified RF power output, it may be harmful for the transceiver over long transmission periods while using continuous (high) duty cycle modes, such as digital, FM and image modes. While a transmitter may be rated at 100 or 200 W, the relevant question "for how long?" should always come to mind. Pay attention to any specified duty cycle mentioned on the manufacturer's specifications side of the Product Review data table. If you're interested in RTTY, FM or other high duty cycle modes, carefully check the manual for any further information.

If no duty cycle is specified, then consider lowering the power output to 30 to 40 W for a 100 W transceiver during high duty cycle mode transmissions. Although modern transceivers incorporate protection circuitry, if you're not sure of your transceiver's ratings, reducing power may prevent overheating

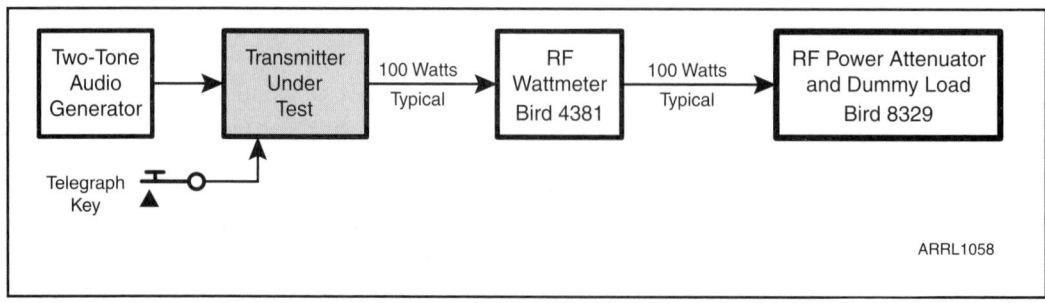

Figure 13.2 — Power output, spurious signal, and harmonic suppression test setup.

Turn Down the Power

It's good practice to operate a transceiver at less than its maximum RF power output if signals are strong. Remember — using the minimum RF power output to maintain communication is required by the FCC rules.

As explained in Chapter 12, every S unit change correlates to a 6 dB change. If an operator lowers power from 100 to 50 W, the operator on the receive end will notice only a ½ S unit drop of signal strength. Dropping the RF power output from 100 to 10 W represents a 10 dB drop; less than two S units. Dropping the power to a lower level is not only good for the final amplifier stage, it's also courteous to your fellow amateur who may be trying to listen to a weak signal on an adjacent frequency.

and lengthen the life of the RF power amplifier components. (See the sidebar, "Turn Down the Power.")

13.2 Emission Standards, a Must

A spurious signal is any signal (unwanted) other than the intended (fundamental) transmitted signal. Any oscillator will create unwanted products, such as spurs (spurious emissions) and harmonics (multiples of the fundamental frequency). Some consider harmonics to be spurious emissions, but for the purposes of this book we'll separate the two.

Spurs are usually found near or around the fundamental frequency. They can be reduced with appropriate circuit design and, to some extent, with filtering. Harmonics are suppressed to an acceptable level with the use of band pass filters. An "acceptable level" is one that will not create interference on two times the fundamental frequency (second harmonic), three times the fundamental frequency (third harmonic) and so on.

The single most important standard your HF transceiver (or any transceiver operating in the ham bands) must meet is the FCC Part 97 rules for emissions. According to Part 97.307d and 97.307e, an HF transceiver's emissions must have at least 43 dB of spurious emission and harmonic

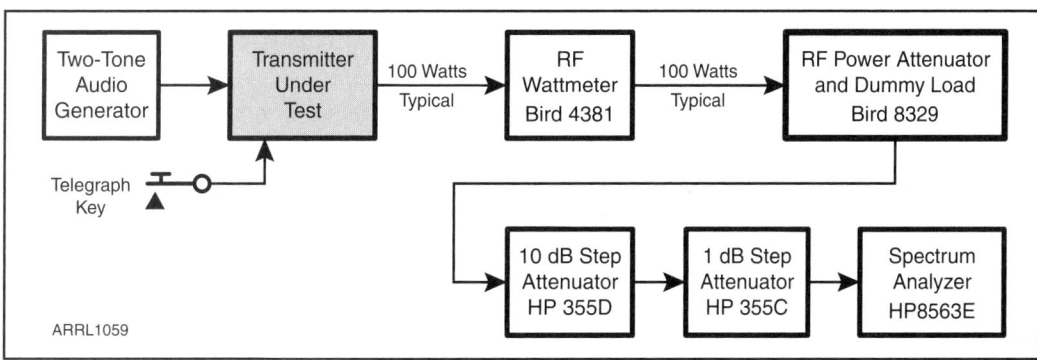

Figure 13.3 — Spurious signal and harmonic test setup.

suppression below 30 MHz. For 30 MHz through 225 MHz, the spurious and harmonic suppression must be 60 dB.

Figure 13.3 shows the spurious signal and harmonic suppression test setup. It builds on the power output test setup, adding 10 dB and 1 dB step attenuators along with a spectrum analyzer. Using a 30 dB power attenuator and a step attenuator, a transmitter's RF power output is fed into the spectrum analyzer at a 1 mW level (sometimes lower for more critical testing). The analyzer is set to sweep a portion of spectrum where any spurious emissions may be transmitted. The purpose of this test is to measure the level of spurious and harmonic emissions. Although the Lab measures spectral purity on all bands where the transmitter operates, only the worst case band is reported. In the example transceiver, the worst band was 10 meters with 62 dB of suppression, nearly 20 dB better than required by FCC rules. All other bands were even better.

Testing for FCC emission standard compliance is the first step taken at the ARRL Lab when evaluating a transceiver for Product Review. In addition, a transceiver's compliance with emission standards is required for a manufacturer or distributor to advertise in *QST*. If a transceiver doesn't pass, we stop testing and notify the manufacturer or distributor.

The goal is to have the lowest possible spurious emissions. Amateur equipment must be held to high standards so our radio service does not interfere with other radio services, and most manufacturers have done a wonderful job in this department. Some inexpensive handheld FM transceivers do not comply with FCC emission standards, however, and you will not see such devices advertised or reviewed in *QST*.

13.3 Spectral Plots

Let's take a look at a few spectral plots of emissions from HF transmitters. The top of each of the following plots represents a 1 mW (0 dBm) fundamental signal and is used as reference for all unwanted emission; the X axis is the frequency, in MHz. Harmonics are always on multiples of the fundamental frequency, and the rest of the unwanted signals are spurs.

Figure 13.4 is a plot from a QRP transmitter kit and shows a fundamental

Figure 13.4 — QRP transmitter kit plot showing a fundamental signal at 7 MHz, second harmonic at 14 MHz, and third harmonic at 21 MHz.

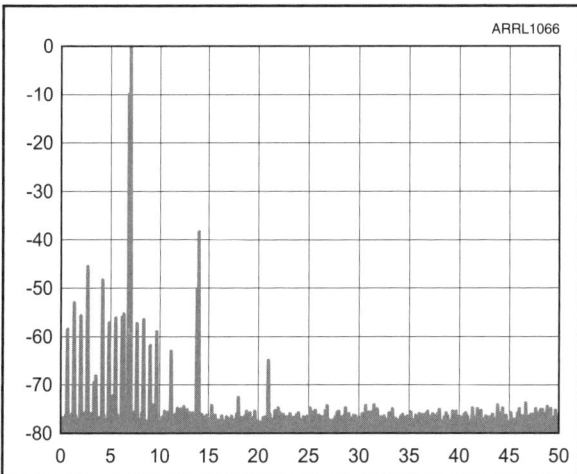

Figure 13.5 — Homebrew transceiver kit plot showing the second harmonic with less than the 43 dB of suppression needed. A rewound output filter with the correct number of turns solved the problem. Several other spurious emissions are seen, but they are within the Part 97 FCC requirement.

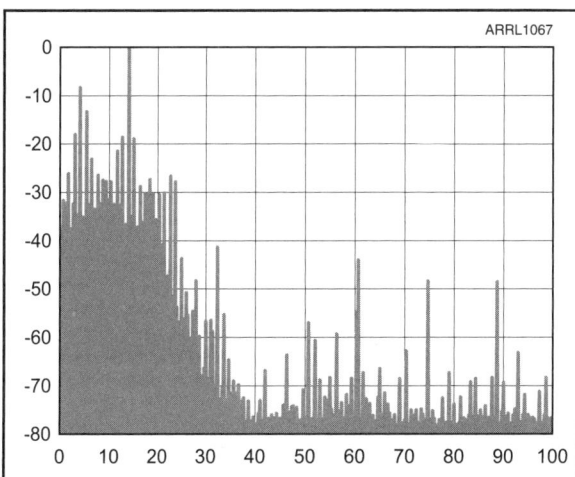

Figure 13.6 — Back to the drawing board for this 20 meter transceiver. It has many spurious emissions that are not even close to meeting FCC Part 97 requirements.

signal of 7 MHz, the second harmonic at 14 MHz, and the third harmonic at 21 MHz. The second harmonic is 45 dB down, close to the needed 43 dB suppression but good enough for FCC rules compliance. Most modern HF transceivers greatly exceed the needed suppression on HF and meet or exceed the suppression level needed at 50 MHz and up. Some low-cost transceivers or kit transceivers may have a suppression figure of 45 to 50 dB on HF.

It is important to remember that our original amateur bands were arranged to be harmonically related. The second harmonic of the 1.8 to 2.0 MHz band (160 meters), for instance, is 3.6 to 4 MHz. If a 160 meter transmitter had a strong second harmonic, at least it would fall in a ham band and not in a region occupied by another radio service.

Figure 13.5 shows the output of a home built 40 meter transceiver kit. The second harmonic at 14 MHz has about 38 dB suppression, less than the 43 dB of needed to comply with FCC rules. The builder eventually rewound the output filter with the correct number of turns on the toroid, which reduced the second harmonic strength to below the required level. The rest of the blips above and below the fundamental are spurious emissions, one of which is close to 43 dB below the fundamental frequency.

Figure 13.6 shows a truly awful transmitter spectrum from a 20 meter transmitter tested in the ARRL Lab. Along with the harmonics, it generates an incredible number of high-level spurs throughout the HF spectrum and higher. Compare this plot with the output from the transmitter in Figure 13.4.

13.4 A Real Life Case

Even though a station's transceiver meets FCC emission standards, it's still possible that the second or third harmonic will be heard. I once received a phone call from a ham who had gotten a report that his station had a harmonic that was heard on the 80 meter band. The gentleman in question was operating the legal limit (1500 W) on 160 meters during the ARRL 160 Meter Contest and was heard as an S7 signal, interfering with a QSO on 80 meters. He did a rough measurement and found his total harmonic suppression was 50 dB — 7 dB more than required by FCC Rules Part 97 Emission Standards. Indeed, he was well within standards, but I explained it this way:

1500 W = 0 dB suppression, S9 + 38 dB
150 W = 10 dB suppression, S9 + 28 dB
15 W = 20 dB suppression, S9 + 18 dB
1.5 W = 30 dB suppression, S9 + 8 dB
150 mW = 40 dB suppression, almost S9
15 mW = 50 dB suppression, S7

Of course, antenna gain factors into this example and the S meter signal strength may vary, but you can see that even with 50 dB of suppression the signal will still be strong enough to be heard. As most avid low power (QRP) operators know, even a few milliwatts radiated from an antenna can be heard at a distance via sky wave. Ed Hare, W1RFI, the ARRL laboratory manager, has worked most states on HF using less than 1 mW!

Chapter 14

Carrier and Unwanted Sideband Suppression

A single sideband signal starts out with two sidebands and a carrier. How well does a transmitter get rid of the unwanted components?

Manufacturer's Specifications	Measured in the ARRL Lab
SSB carrier suppression: At least 60 dB.	>70 dB.
Undesired sideband suppression: At least 60 dB.	>70 dB.

14.1 Transmitting a Clean Signal

An AM signal consists of a carrier and its upper sideband (USB) and its lower sideband (LSB), which are used to convey audio (speech). For single sideband (SSB), the carrier and unwanted sideband must be suppressed. For instance, when transmitting in upper sideband, the lower sideband and the carrier are removed to create a single sideband transmission.

The lower sideband and the carrier are not needed for transmission, but they are not totally eliminated by the transmitter's circuitry. Rather, they are suppressed to a very low level. Any leftover, unwanted sideband or carrier can cause interference to another station that is close by. In an ideal world, there could be an SSB communication simultaneously happening every 2.5 kHz up and down the band without interference, but that is not always be possible due to transmitter performance and operator error (we'll get into that later).

14.2 Making the Measurements

The levels of these unwanted signal components are measured by a carrier and unwanted sideband suppression test. The measurement is made using a spectrum analyzer and injecting a single audio tone into the microphone jack while transmitting in the USB mode. **Figure 14.1** shows the carrier and unwanted sideband suppression test setup, which is the same as the one we used for the harmonic and spurious emission tests in the previous chapter.

In this case, the spectrum analyzer is set for a narrower frequency span, 20 kHz, to display the desired transmitted USB audio tone, the unwanted LSB audio tone, and the suppressed carrier. The desired transmitted USB audio tone creates an RF signal level that is used as reference. The difference between the desired USB signal and the undesired LSB signal is measured in dB. The test is repeated with the desired transmission made on LSB, and the undesired USB signal and carrier level are measured. For a properly operating SSB transmitter, the unwanted sideband and carrier suppression should be the same in the USB and LSB mode.

Figure 14.2 shows a signal generated by an audio tone on USB. The carrier, dead center of this plot, is just out of the noise at 70 dB below the audio tone. The carrier suppression is 70 dB. The unwanted (lower) sideband signal is the small blip left of center and is 64 dB down. Therefore, the unwanted sideband suppression is 64 dB.

The established manufacturers offer excellent carrier and unwanted sideband suppression in modern transceivers, and we have measured greater than 70 dB for each at the ARRL Laboratory. I consider 60 dB to be good suppression and 50 dB to be the cutoff for acceptable.

Remember the example I used in the previous chapter? Would you want

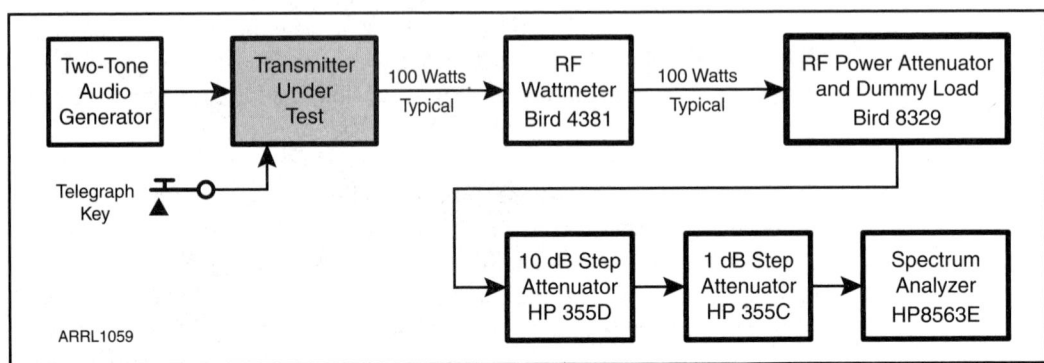

Figure 14.1 — Carrier and unwanted sideband suppression test setup.

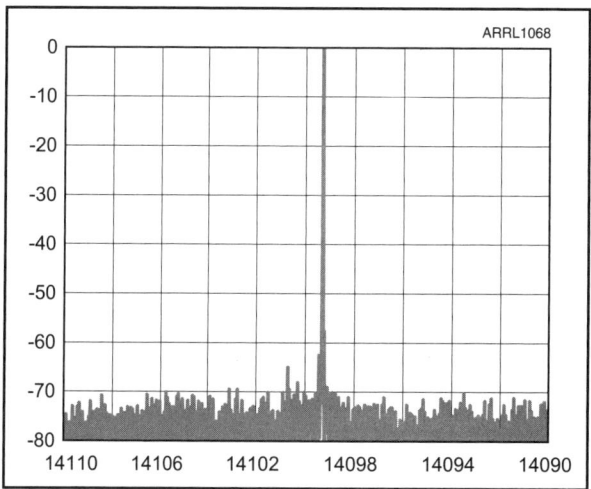

Figure 14.2 — Plot showing an upper sideband (USB) transmission on 20 meters. The strong signal just to the right of center is generated by a single audio tone on upper sideband (USB). The carrier is dead center at 70 dB below the USB tone, and the LSB signal just to the left of center, at 64 dB below the USB tone.

your legal-limit station to transmit lower sideband or a carrier at the 15 mW level? I hope that all radio amateurs aspire to transmit the cleanest signal possible and to set a good example in their use of technology. For radio amateurs in the market for an HF transceiver, the unwanted sideband and carrier suppression should be as high as possible, but at least 50 dB for each figure.

Chapter 15

Two-Tone Intermodulation Distortion Products — Transmitted IMD

Intermodulation distortion (IMD) products can cause your transmitted SSB signal to "splatter" and cause interference to stations above and below your frequency. How well are IMD products suppressed?

Manufacturer's Specifications
Third-order intermodulation distortion (IMD) products: –31 dB @ 14 MHz, 100 W PEP.

Measured in the ARRL Lab
HF, 100 W PEP, 3rd/5th/7th/9th order:
 –27/–40/–42/–52 dB (worst case, 10 m);
 >–31/>–40/>–45/>–52 (typical).
50 MHz, 100 W PEP: –32/–38/–45/–61 dB

15.1 What Is Transmitted IMD?

Any amplifier will exhibit some unwanted mixing products during operation. In an amateur transmitter there is amplification in the audio stage (microphone preamp) that is used to modulate the driving stage(s) and the final amplifier stage, so there are several opportunities for generation of unwanted products in a transmitted signal.

The most measureable unwanted products are *odd-order products*, similar to those observed with two strong adjacent received signals that create receiver third-order IMD products as discussed in Chapter 6. The third, fifth, seventh, ninth and higher order transmitted products add to the distortion of the transmitted signal. If these levels are high enough, they will cause interference (splatter) and simultaneously reduce the transmitter's effectiveness.

15.2 How Transmitted IMD Is Measured

The purpose of the transmitted IMD test is to determine the odd-order products that are transmitted during normal operation of an SSB transmitter. The measurement is made using an audio generator to generate two audio tones, free from odd-order products, which are injected into the microphone jack of the transmitter under test. The transmitter under test is brought up to full peak envelope power (PEP) RF output by adjusting both power output level and microphone gain while monitoring the transmitter's ALC meter. **Figure 15.1** shows the two-tone IMD test setup. It's the same one used in the previous two chapters.

With just enough two-tone audio injected to activate the ALC circuitry, a spectrum analyzer is used to measure the transmitted IMD. The transmitter is set for rated PEP output. Measurements are made on each band, and the worst-case band is reported along with typical measurements from the other bands.

This measurement is made with each audio tone set to 6 dB below the top of the scale (the top of the scale mathematically works out to the full PEP RF output power). **Figure 15.2** shows the spectrum analyzer display of a two-tone IMD test for a good transmitter. Note that the unwanted IMD products appear on each side of the two audio tones and extend many kHz from the center of the signal (the ninth-order products are about 5 kHz above and below center). During voice operation, as an operator speaks into the microphone, the IMD products are transmitted within and on both sides of the center frequency. If they are strong enough, they can cause splatter for an amateur listening nearby.

For today's transceivers, a third-order product measurement of

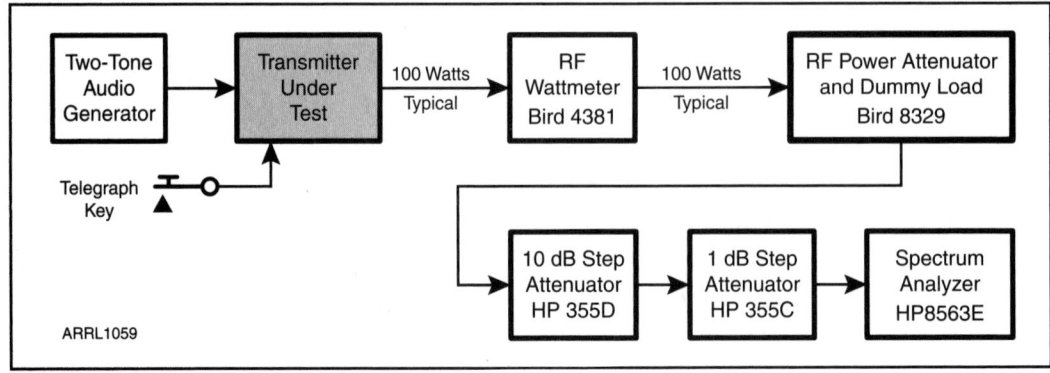

Figure 15.1 — Two-tone intermodulation distortion test setup.

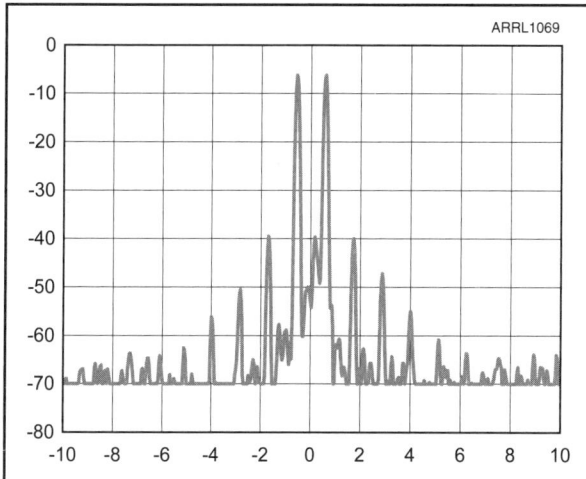

Figure 15.2 — Two-tone IMD test results observed on a spectrum analyzer. This is a good transmitter — the 3rd order products are close to 40 dB below peak power and the higher order products grow progressively weaker.

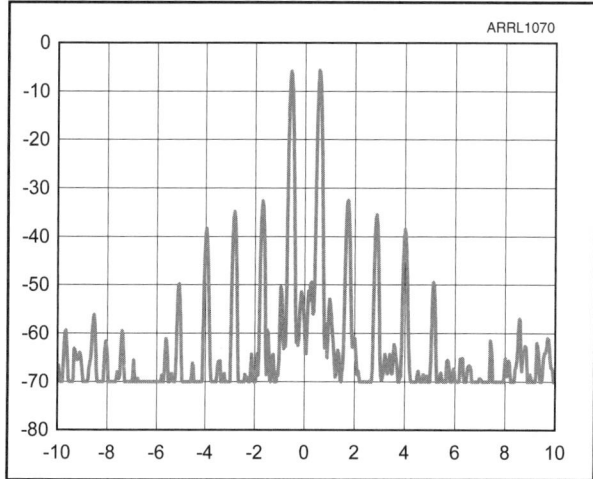

Figure 15.3 — In this signal, third-order IMD products are okay at −32 dB but the higher order products are too strong.

30 dB below PEP is typical, 35 dB is considered good and 25 dB below PEP is mediocre. A clean signal with lowest possible IMD products should be a key consideration for anyone choosing an HF transceiver.

15.3 Examples of Transmitter IMD

Figure 15.2 shows what all manufacturers should strive for: third-order products close to 40 dB below peak power. Voice transmissions from the transceiver that created this plot have low distortion products and will noticeably stand out because of their clean audio.

Figure 15.3 shows a plot of transmitted IMD levels of a modern transceiver with poor IMD characteristics. Although the third-order products are 32 dB below peak power — not too bad — the fifth-order products are only 35 dB below peak power and the seventh-order products are only 38 dB down. As you can see, ninth-order products about 5 kHz away from the center channel are only 50 dB below peak power, which is higher than desirable. This signal will sound "wide" on the air.

Figure 15.4 shows a transmitter with a different problem. The third-order products are higher, nearly 25 dB below peak power. The fifth, seventh and ninth-order products are all significantly better than the transmitter shown in Figure 15.3. With third-order products less than 25 dB below peak power, noticeable distortion can be heard on the air and transmitted audio degrades. If I see a poor IMD reading in the lab, I pass the information along to the manufacturer early in the review process.

Two-Tone Intermodulation Distortion Products — Transmitted IMD

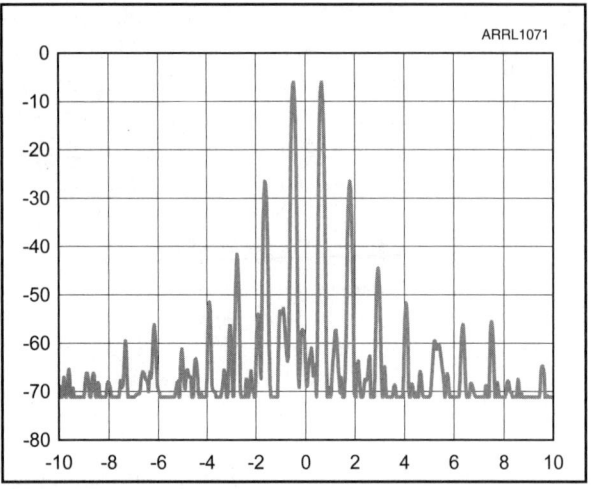

Figure 15.4 — In this signal, the higher order products are okay but third-order products are only about 25 dB below peak power.

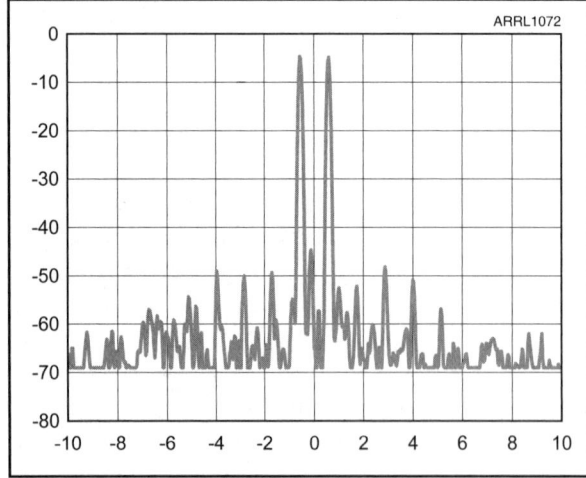

Figure 15.5 — This very clean signal (third, fifth and seventh-order products all about –50 dB) was generated by a Class A transmitter operating at 75 W output.

Considerably lower IMD products are transmitted with Class A modulation, available as a feature on a handful of transceivers. Class A transmitters can exhibit third-order products at or below the 40 dB below PEP level. While Class A modulation produces a very clean transmitted signal, the circuitry of Class A modulation requires more current from the power supply and places more demand on the final amplifier, so the power output must be reduced. This feature also increases the cost of the transceiver, but some amateurs feel that the extra performance is worth the extra expense.

Figure 15.5 is a plot of a transceiver using Class A modulation. The transceiver is at its full Class A power output (75 W), with the microphone gain adjusted just enough to make full power. The ALC level at this point is about midway in the ALC window. The third, fifth and seventh-order products are all about 50 dB down. This is exceptionally clean modulation.

15.4 Watch the Automatic Level Control (ALC) Level!

One topic I stress with new radio amateurs is monitoring the ALC meter while transmitting in the SSB mode, especially if they're using a computer sound card that creates audio tones that are injected into the transceiver's audio chain for digital transmissions.

It's human nature to want the strongest possible transmitted signal (QRP op exception acknowledged). While transmitting, we want our external power/SWR meter to read as much power as our station can put out.

The problem is that most wattmeters do not react fast enough to show actual PEP power, so the operator turns up the microphone gain and/or the speech processor level and calls that rare DX station, trying to be heard over the din of everyone else. But with both microphone and speech processor turned up, the ALC can be easily exceeded. When this happens, the transmitted IMD increases and can become quite bad, causing nasty splatter on the band. A wide signal that splatters means that the transmitter is not concentrating all the power it can generate to the signal to be detected at the other end. We've mentioned the receiver improvements made by filters, but there's no filter today that can remove the splatter caused by a radio amateur who exceeds the ALC range while transmitting.

   Minimal ALC must be used while transmitting audio tones used for digital operation. Remember that ALC circuitry is designed for speech, not audio tones. That is why it's important to adjust the audio control of the transmitter just enough to visually detect a small amount of ALC indication on the meter. If more ALC is indicated or the ALC limit is exceeded, the transmitted IMD greatly increases and the digital signal will be wider on the air and appear wider on a waterfall display at the other end. Many have complained to me that the manufacturers don't do enough to reduce IMD products. While transmitters could be better, it's often operator error that causes splatter and wide digital signals.

Chapter 16

CW Keying Waveform

What will a transmitted CW signal sound like on the air? In the Lab we make several measurements of keying characteristics, as described in this chapter and the next. First up, how does the CW signal sound on the air? Is it a nice smooth tone?

Manufacturer's Specifications	Measured in the ARRL Lab
CW keying characteristics: Not specified.	See Figures 1 and 2.

16.1 Keying Waveform Test

The shape of a transmitter's keying waveform impacts the quality and effectiveness of its transmission. *QST* publishes the keying waveform of each HF transceiver it reviews. The purpose of the keying waveform test is to determine if the waveform has any characteristics that will have a negative impact on the quality or tone of the transmitted signal.

In some cases, the transmitter sends a short high-power spike at the beginning of a transmission. Called *overshoot*, this can activate a power amplifier's protection circuitry. Overshoot can be identified by examining the keying waveform.

Figure 16.1 is a block diagram of the CW keying waveform test setup. The keying waveform test is performed by externally keying the transmitter at a rate of 60 words per minute (WPM) and capturing the first two transmitted "dits" sent with a storage oscilloscope. A power attenuator and step attenuator are used to reduce the transmitted RF output to a safe level for the oscilloscope.

16.2 Examples of Keying Waveforms

Let's look at a few examples of keying waveforms and how they affect the transmitted signal. Our test is performed with the transceiver in the full

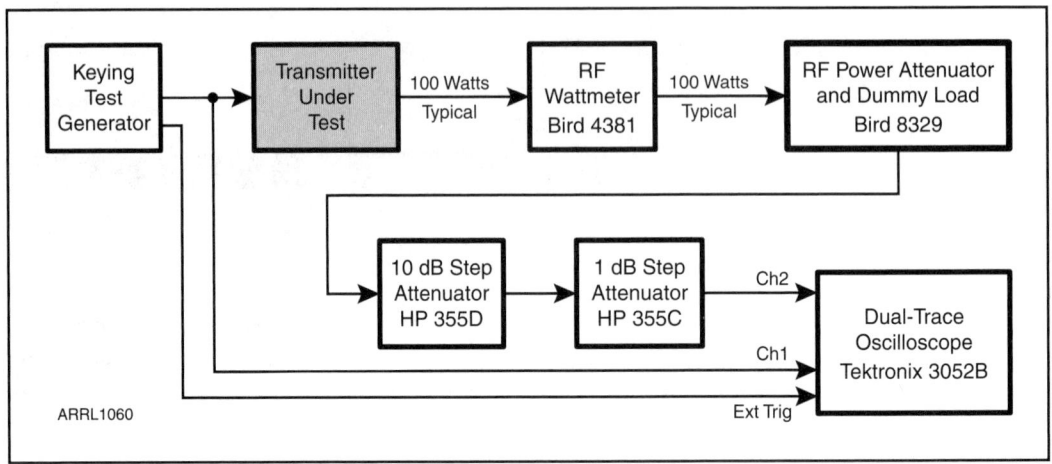

Figure 16.1 — CW keying waveform test block diagram.

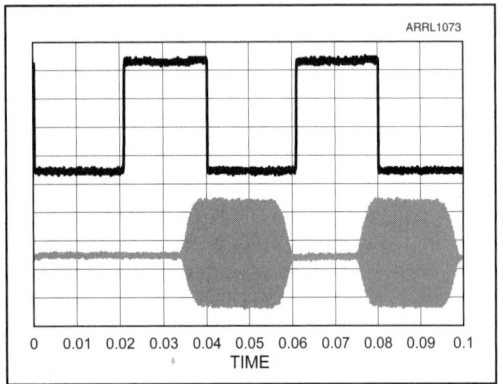

Figure 16.2 — An ideal waveform, with the dits the same shape and of equal duration.

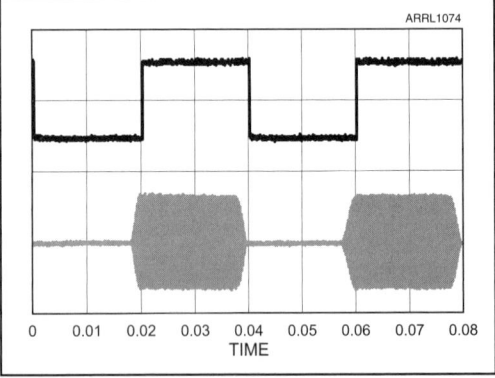

Figure 16.3 — Here we have good waveform shaping and about half the waveform delay as that shown in Figure 16.2.

break-in (QSK) mode, if available. Each keying waveform plot shows two waveforms. The top waveform is the external keying generator, which acts like a Morse code straight key closing and opening; the low part of the waveform is key down, the high part is key up. The bottom waveform is the transmitter's keying waveform. The first key down action after the keying generator occurs at time zero and is seen at the very left of the plot. The plot shown in **Figure 16.2** is an ideal keying waveform shape. The first dit is the same shape as the second one and the duration of each dit is equal. The

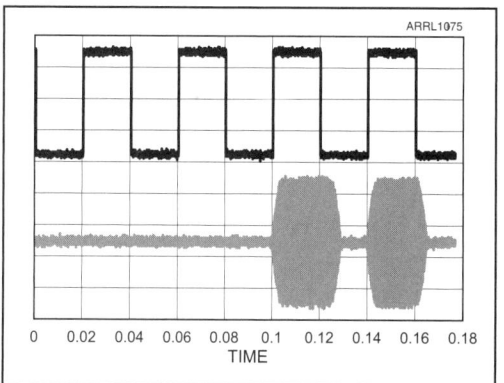

Figure 16.4 — This is also good waveform shaping, but the first dit is delayed considerably.

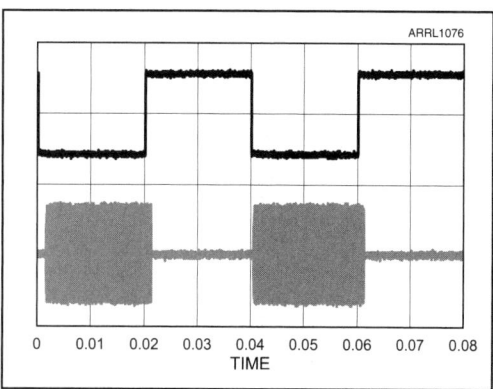

Figure 16.5 — This very abrupt, nearly vertical rise to the keying waveform indicates hard keying.

beginning of the waveform starts 35 ms after the transmitter is first keyed, which is acceptable, though a faster reaction is preferable. This particular plot is from an SDR transceiver.

Figure 16.3, from a traditional analog transmitter, shows good waveform shaping and about half the waveform delay as Figure 16.2. The shaping of the first and second dit are quite similar and of the same duration, providing the receiving operator with a good signal.

Figure 16.4 shows a plot of the longer duration that was needed to capture the first two dits of an SDR transceiver. It shows good waveform shaping, but the first dit is delayed considerably and does not occur until 100 ms after the transmitter is first keyed. While not a problem for the operator at the receiving end, such a delay can be maddening for the sending operator if the signal is monitored off the air.

Figure 16.5 shows very little transmitter delay, but the keying waveform has a very abrupt, nearly vertical rise on the leading edge. This is known as *hard keying*. It is audible, sounds abrupt, and can cause fatigue to the person at the other end. Worse yet, this abrupt waveform with its near vertical rise and sharp edges will cause the transmitted bandwidth to increase (we'll learn more about that in the next chapter). Over the air, hard keying can sound like a clicking noise (key clicks) on or above and below the transmitted frequency. This is mainly caused by a receiver's noise blanker when it is engaged, rather than by the transmitter itself. A gradual rise time of each waveform, on the other hand, is known as soft keying. Soft keying is difficult to hear under less than ideal reception.

Figure 16.6 was not made for publication, but to inform a manufacturer of a problem with the keying waveform at half the specified RF power output.

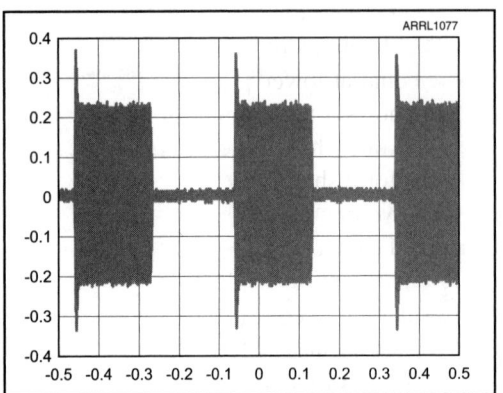

Figure 16.6 — A spike at the top of the leading edge of each dit indicates an unwanted product that causes key clicks.

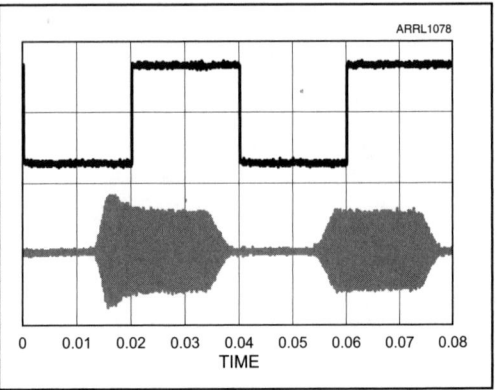

Figure 16.7 — The bump at the first dit is a sign of transceiver overshoot.

The ARRL laboratory routinely checks for any keying waveform issues at half power and at full power. Here, at the top of the leading edge of each dit, we can see a spike that is an unwanted product that causes key clicks. If the key clicks are severe enough (as they are here), annoying clicking sounds can be heard up and down the band. This manufacturer was happy to be informed of the issue and made efforts to correct it. Back in my Novice days, around 40 years ago, key clicks were heard occasionally on the ham bands, and the operator of the offending transmitter quickly got a letter from an ARRL Official Observer, or even a warning letter from the FCC. Today's transmitters are a lot better and we don't often find key clicks such as these.

Looking at **Figure 16.7** we see a significant bump at the beginning of the first dit. This indicates what is known as overshoot, and it occurs when a transmitter's circuitry does not control the RF output level quickly enough. Audibly, the first dit will sound like hard keying since the RF output at the peak of the bump is greater, by about 20% in this case. While this will not cause an over-the-air key click, it will cause some linear power amplifiers to trip off. At the time of the bump, the power amplifier is momentarily overdriven, which is detected by the amplifier's protection circuit, and it will shut down to protect itself. This was another instance where we notified the manufacturer, who was grateful to learn of the issue. Efforts were made to correct the problem, which reduced and flattened out the bump to a more acceptable waveform shape.

The plot in **Figure 16.8** shows that the first dit is half as long as the following dits sent. This undesired effect can cause confusion to the receiving operator if high CW speeds are used. If a dash (dah) is the first component

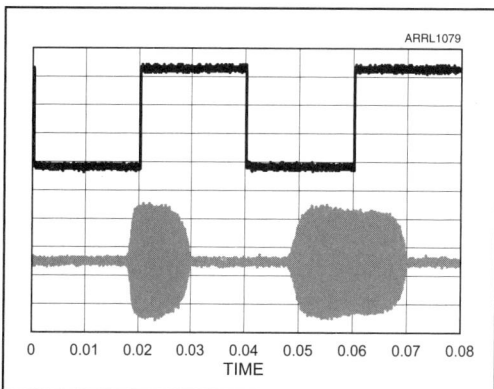

Figure 16.8 — The first dit is truncated to about half the length of the second dit. This can cause confusion at the receiving end at high speeds — dits may disappear, and shortened dahs may sound like dits.

of a Morse letter sent, it may be mistaken for a dit. The durations of the first and second dit become equal at lower code speeds. Note that this plot shows a fair amount of overshoot on the second dit; while it is tolerable, it's not an ideal waveform shape.

All CW operators who wish to be considerate of other amateurs using the band should pay attention to the keying waveform when evaluating an amateur transceiver. Serious contesters should also pay attention to the keying waveform so they can choose a transceiver that will increase the effectiveness of the transmitted signal during crowded band conditions. Some mid- and top-end transceivers have a menu setting for adjusting the CW rise time, allowing the user to adjust for harder or softer keying.

Chapter 17

CW Keying Sidebands

Keying sidebands can cause a CW signal to occupy more spectrum than it should, resulting in interference to nearby stations. Here's how we test for that.

Manufacturer's Specifications
CW keying characteristics: Not specified.

Measured in the ARRL Lab
See Figures 1 and 2.

17.1 Data Rate versus Bandwidth

We tend to think of a CW signal as occupying one discrete frequency, but that is not the case. While a steady, continuous unmodulated carrier may appear to occupy one discrete frequency, it will always have some noise on each side of the carrier, generated mostly by a transmitter's oscillator (more on that in the next chapter).

As the sending speed of a CW transmitter is increased, the occupied bandwidth is also increased. As with other modes, the greater the information (data) rate that is sent, the greater the bandwidth that is needed. A transceiver's keying waveform shaping can have an impact on the transmitted bandwidth, and the purpose of the CW sideband test is to observe and measure that bandwidth.

Figure 17.1 shows the keying sideband test setup. The test is performed by observing the transmitter's output in the CW mode while sending a string of "dits" at 60 WPM. The RF output level of the transceiver is reduced to a safe (1 mW) level with power and step attenuators. The transmitter is set to 14.020 MHz, and the spectrum ±5 kHz is observed using a spectrum analyzer.

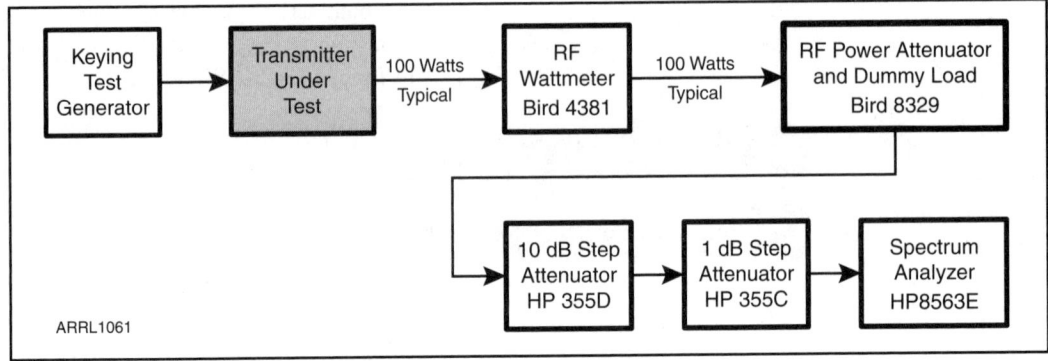

Figure 17.1 — Test setup for measuring CW keying sidebands.

17.2 Examples of CW Sidebands

Let's take a look at a few examples. Each of the following plots represents many "snapshots" taken over time in order to provide sufficient data points for meaningful results. The RF signals generated on each side of the center signal are known as CW sidebands and appear 5 kHz above and below the center frequency.

Figure 17.2 shows the CW sidebands from the transmitter that had the keying waveform shape shown in Figure 16.3 in the last chapter. At 1 kHz either side of the center frequency, the CW sidebands are about 60 dB below the carrier. On the air, if signals were strong, the CW sideband noise would raise the noise floor in the receiver of an operator 1 kHz away. That would mask weak signals the adjacent station is trying to hear.

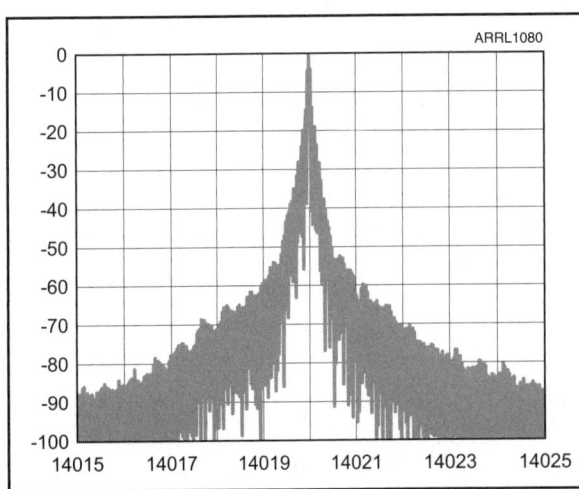

Figure 17.2 — CW keying sidebands from the transmitter with the keying waveform seen in Figure 16.3. This plot is fairly typical of modern transceivers.

Figure 17.3 is an example of a transmitted signal with very good CW sidebands. At 1 kHz from the center frequency the keying sidebands are down by about 75 dB, 15 dB better than the transmitter shown in Figure 17.2. This transceiver will generate much less interference to adjacent users at high code speeds.

In **Figure 17.4** we have a plot of an inexpensively manufactured QRP transceiver with rather wide CW

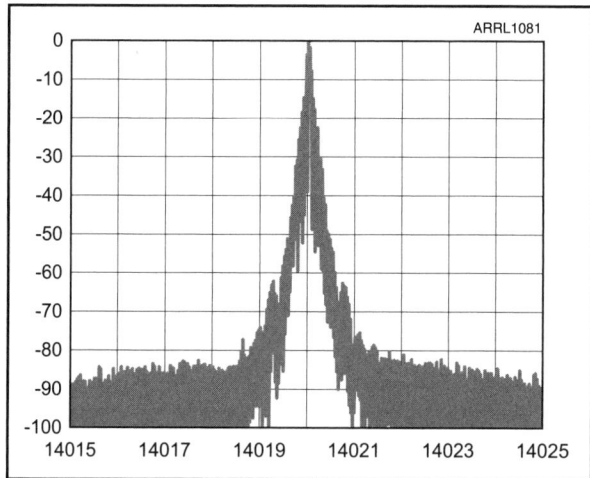

Figure 17.3 — This plot shows very good CW keying sidebands. This transceiver will generate much less interference to adjacent users at high code speeds.

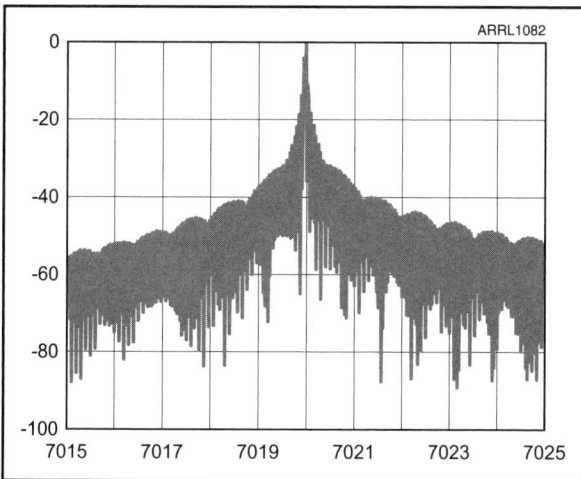

Figure 17.4 — This inexpensively manufactured QRP transceiver has rather wide CW sidebands. The near vertical rise and fall time of the keying waveform is the main cause of this.

sidebands. Each sideband is more than 20 dB stronger than the signal shown in Figure 17.2, and the sidebands are very still strong 5 kHz away from center. This plot was taken with the transceiver that demonstrated the hard keying characteristics (Figure 16.5) discussed in the previous chapter. The near vertical rise and fall time of the keying waveform is the main cause of the wide and higher-than-normal CW sidebands. While these sidebands might be borderline for a QRP transmitter (5 W or less), they are lower at slower sending speeds. Still, I would not want to risk making my noisy signal even worse for others on the band by using an RF power amplifier to boost RF power output to the 100 W level.

When in the market for an amateur transceiver, look for reasonable CW sidebands if CW is a mode you wish to pursue. Other CW operators will appreciate your choice of transceiver.

Chapter 18

Composite Noise Test

Transmitters generate noise in the oscillator and amplifier stages, and that noise is present at the transmitter output. This is called composite noise, and if it is too strong it can cause interference to other stations on the band.

Manufacturer's Specifications	Measured in the ARRL Lab
Composite transmitted noise: Not specified.	See Figure 3.

18.1 What Is Composite Noise?

One transmitter performance level that cannot be adjusted in any way is the noise generated from the oscillator and additional transmitter stages. All oscillators generate noise, such as a local oscillator's sideband noise, as discussed in the Chapter 5. The higher the quality of an oscillator, however, the lower the noise it generates. The noise is a composite of both amplitude and phase noise elements. The purpose of the composite noise test is to determine the amount of noise generated 100 Hz to 1 MHz away from the oscillator carrier.

18.2 The Measurement

The measurement is made with a specialized composite noise test set, composed of a special computer-controlled comparator, a spectrum analyzer, and a signal analyzer. A power attenuator and a step attenuator

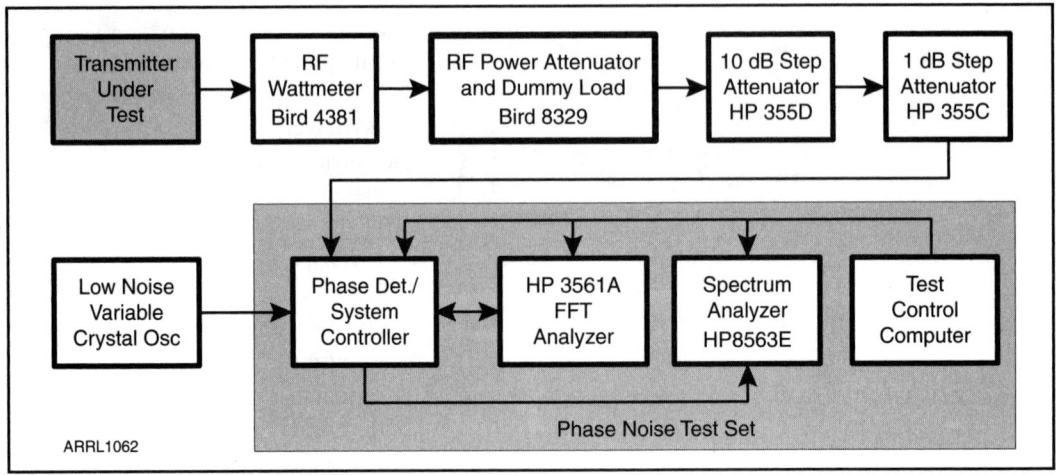

Figure 18.1 — Composite noise test setup.

are used to reduce the RF output of the transmitter under test to a usable level (1 to 5 mW). This signal is fed into one of the comparator's two ports. A high-quality test oscillator with composite noise far below the test transmitter's oscillator is used for comparison and is fed to the other port. **Figure 18.1** shows the composite noise test setup. The carriers are locked together and cancelled out by the comparator, leaving only the transmitter's composite noise as an RF signal, which is plotted by the computer's software. Radio amateurs should consider composite noise if they wish to be considerate to other users of the band; the lower composite noise, the better.

18.3 Examples of Composite Noise

Let's take a look at a few composite noise plots made at the ARRL Laboratory. Each plot shows the noise response versus frequency. The Y axis shows the noise in dB, relative to the missing carrier set to a reference level of 0 dBm.

In the example shown in **Figure 18.2**, the transceiver's composite noise is at, or better than, 100 dB below the carrier

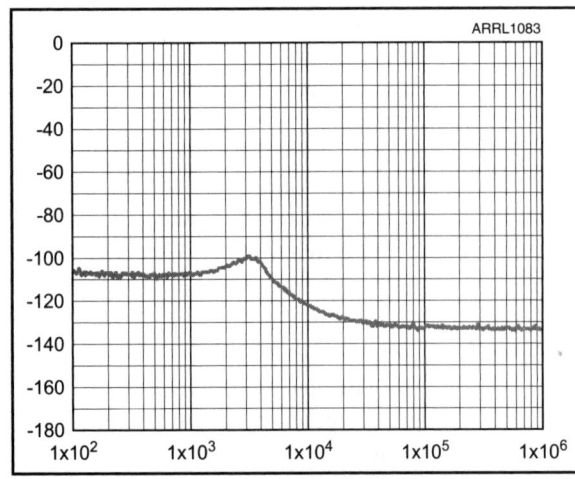

Figure 18.2 — This plots shows the composite noise at or below 100 dB below the carrier between 100 Hz (1×10^2) and 1 MHz (1×10^6) away from the carrier. This transmitter will generate noise 100 dB below its peak power about 3 kHz away from the carrier frequency.

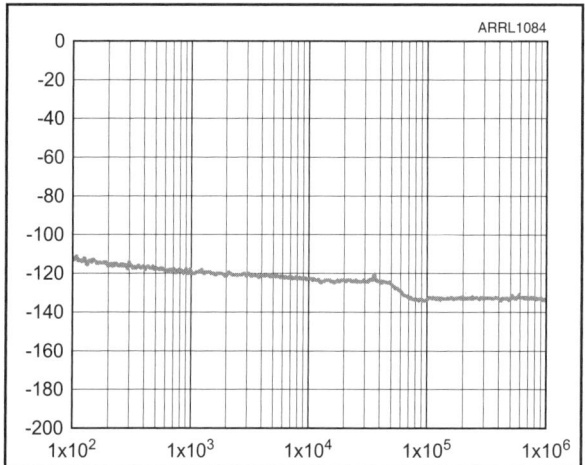

Figure 18.3 — The low composite noise demonstrated here means it's unlikely this transmitter will add to the noise floor unless it is close to a receiver.

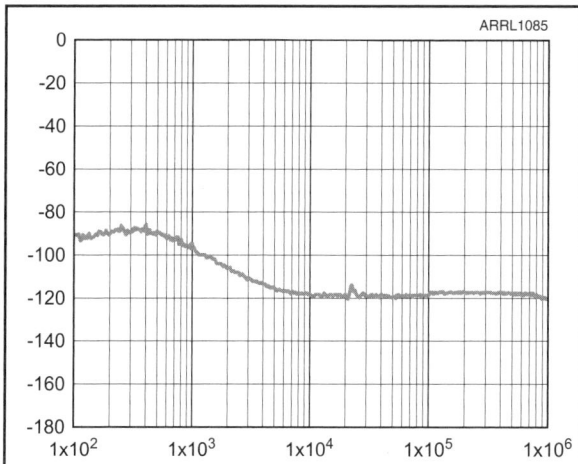

Figure 18.4 — The noise we see peaking higher than 90 dB below the carrier between 300 and 400 Hz away from the carrier means this transmitter may cause interference if its transmitted signal is strong at the receive end.

between 100 Hz and 1 MHz. According to this plot, the transmitter under test will generate noise 100 dB below its peak power about 3 kHz away from the carrier frequency. This is the bump between 1×10^3 (1 kHz) and 1×10^4 (10 kHz).

What effect will this noise 100 dB below the carrier have on other stations using the band? Let's say you're transmitting at 14.020 MHz, using 1500 W output and a good antenna. Another station 3 kHz up the band is trying to hear a weak signal near his receiver's MDS level, say at −135 dBm. If your main signal is strong at his location, say S9 + 40 dB (−33 dBm), the composite noise 3 kHz away will be 100 dB weaker than that, or −133 dBm. That's strong enough to create noise at the receiving station's frequency, possibly masking the very weak signal he's trying to hear. Signals at 40 dB over S9 are at a level often observed on the air when stations are using legal-limit transmitter power and directional antennas.

The plot in **Figure 18.3** shows a transmitter with low composite noise. It's unlikely this transmitter will add to the noise floor unless it is located close to a receiver.

Figure 18.4 shows a transmitter with higher close-in composite noise, peaking higher than 90 dB below the carrier between 300 and 400 Hz away from the carrier. This transmitter has a greater potential to cause interference if its transmitted signal is strong at the receive end.

18.4 Cumulative Effects

In these composite noise examples, we have only considered the effects of one transmitter on one receiver, but what if the entire ham band you're operating on is jammed with strong signals? Because of how far the composite noise extends out from the carrier, we can actually experience a raised noise floor from the cumulative effects of composite noise from each transmitter on that band. It is desirable for manufacturers of amateur transceivers to use the best quality oscillators possible.

Chapter 19

Turnaround Times

How quickly does a transceiver switch from transmit to receive, and back again?

Manufacturer's Specifications	Measured in the ARRL Lab
Transmit-receive turn-around time (PTT release to 50% audio output): Not specified.	S9 signal, AGC fast, 36 ms.
Receive-transmit turn-around time (tx delay): Not specified.	SSB, 34 ms; FM, 30 ms.

19.1 Delays

We may assume that a transmitter will start transmitting as soon as the push-to-talk (PTT) is pressed, but that is not the case. There is always some delay, for example the delay between Morse key closure and RF output discussed in Chapter 16.

A quick turnaround is desirable. The ARRL laboratory considers a turnaround time of less than 35 ms to be satisfactory for both voice and digital work. A turnaround time of greater than 100 ms can result in the first word or syllable of the transmission getting cut off if the operator starts speaking immediately after pressing the PTT button. Consequently, a receiver's recovery time after the PTT button is released is equally important, and the shorter it is, the better. A slow recovery time, with the AGC set to FAST will result in missed information. A rating of 36 ms or less is best for voice and digital work.

Figure 19.1 shows a receive-to-transmit turnaround time test block diagram. The test is performed using a keying test generator for engaging the PTT circuit. An audio tone generator is fed into the transmitter's microphone jack. With the PTT activated, the audio tone modulates the transmitter and

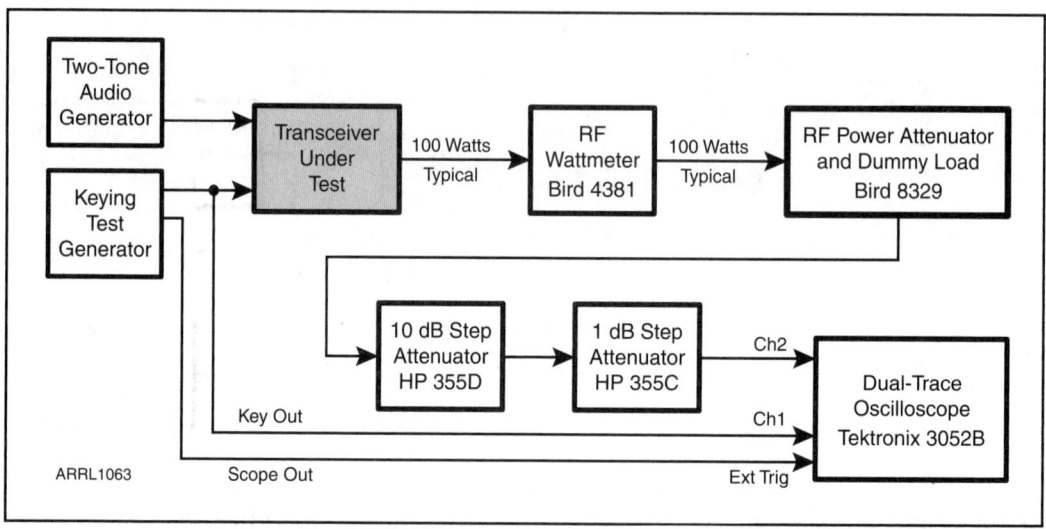

Figure 19.1 — Receive-to-transmit turn-around time test block diagram.

the RF output level is adjusted for 100% of the rated RF power output. The keying generator is adjusted for a long keying cycle of several seconds. Both the keying generator and the transmitter's (attenuated) RF output are fed to a dual trace oscilloscope. With the keying generator turned on, a time delay is observed between the PTT on (key closure) and the transmitter RF output. With the oscilloscope screen set to display both waveforms, the time delay is measured from the key closure point to the point of 50% of the RF output. During the receive-to-transmit test, close observations are made to see if there are any unusual effects, such as RF spikes from the transceiver's PTT circuitry or other distortions with the transmitted audio.

Figure 19.2 shows a transmit-to-receive turnaround time test block diagram. This test is performed using the keying test generator in the same manner as the receive-to-transmit test. For this test, both keying test generator and receiver audio output waveforms are observed. An RF signal generator output is fed, using considerable attenuation, to the antenna jack of the transceiver. Extra attenuation is added between the signal generator and transceiver as a precaution, since RF from transmitter does reach the input jack of the signal generator and it must be no greater than −40 dBm. With the keying test generator turned on, the PTT circuit is engaged and key closure and the absence of speaker audio are observed on a dual-trace oscilloscope. As the keying test generator opens the PTT circuit, the key-open waveform and the audio signal from the receiver are observed and the

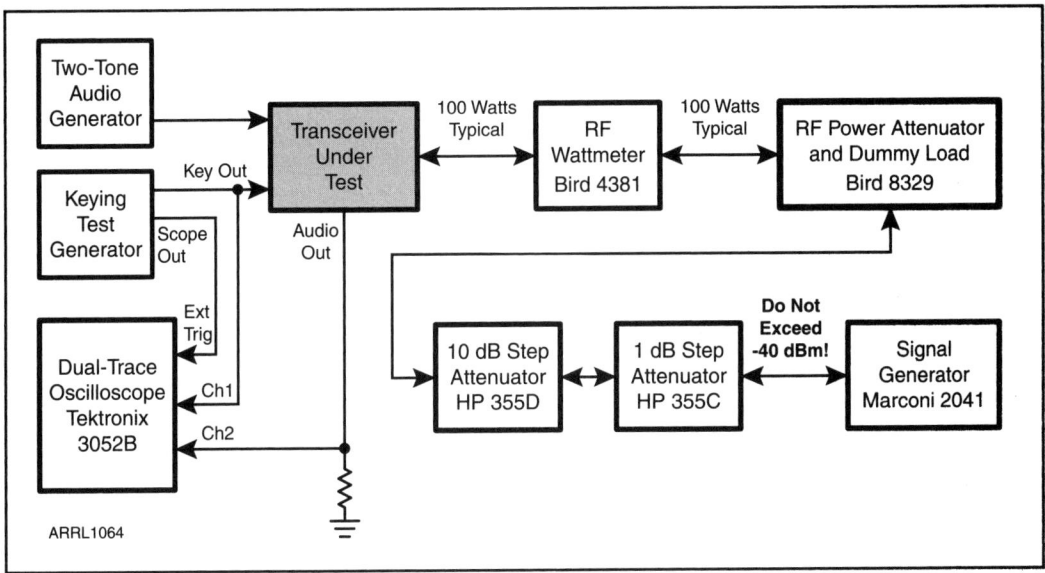

Figure 19.2 — Transmit-to-receive turn-around time test block diagram.

delay time of key-open to 50% of the receive speaker audio is measured. During the transition times, observations are made to look for pops, clicks or other audio distortions in the audio waveform.

19.2 Delays Specific to SDR Transceivers and Receivers

Some SDR transceivers use an outboard personal computer or laptop to control the signal processing software, and this computer's processor speed will affect turnaround times. An older computer with a processor speed of 2 GHz or less can add to turnaround time. When considering the purchase of an SDR transceiver or receiver, find out from the manufacturer (and *QST*'s Product Review) what specifications are needed from a computer for the software to run effectively, with minimum delay times. You may want a little more computing horsepower if the SDR software will also be accessing the Internet and running logging software.

Chapter 20

Other Considerations

Although a transceiver's specifications and performance in the Lab are important, there are several other factors to consider when shopping for a new HF transceiver.

20.1 Beyond Performance

In the previous chapters, I covered the technical side of choosing an HF transceiver. By now the reader should have a general knowledge of the various terms and figures that are presented in *QST*'s Product Review data tables. Although this knowledge is very important to the decision-making process, the user interface must suit the radio amateur's operating style and esthetic tastes. Operating a radio is a sensual experience as well.

20.2 The Look

Many newly licensed amateurs just want to get on the air and talk, just as someone with a brand new driver's license may just want to get on the road and drive. Of course, a lot of drivers don't want to be seen behind the wheel of the family minivan and crave wheels that attract attention. Automobile manufacturers know that "the look" helps sales — so do manufacturers of Amateur Radio equipment. The look can be just as important to some amateurs, even though a flashy transceiver has the same RF power output or receiver performance as other transceivers. The attraction may seem superficial at first, yet the look of a transceiver has major impact on the overall comfort of its user. Nonetheless, see the "Caveat Amateur" sidebar for a bit of a cautionary reminder.

Traditional analog transceivers have some sort of digital display. SDR transceivers may look like traditional analog transceivers with the typical knobs and switches, with some models having one or more colorful graphic displays that provide an advanced visual aid for tuning and other functions. An SDR transceiver can also look like a plain box that can be hidden out of sight. An external computer does the work and the software places every

Caveat Amateur

I admit to occasionally being blinded by the look of a new transceiver. But when that happens, I think back to my youth, when I fell in love with the looks of my first transceiver.

As a 16-year-old Novice, I put in an entire summer at hard manual labor just to save up for my first big rig — a brand new Hallicrafters FPM-300 MK II. I bought this transceiver solely based on its looks, which was all I knew at the time. The 1975 price tag of $695 (almost $3000 in today's dollars) matched my bank account balance. But I was quite smitten with my new rig — until I plugged it in and turned it on! The tuning dial had considerable backlash (it moved by itself when let go), the frequency drifted with temperature, and the non-adjustable CW sidetone had the same tonal characteristics as my mother when she complained about TVI. The stomping of her foot on the floor above started soon after I transmitted on 15 meters, a "side effect" that never occurred with my Heathkit DX-60B transmitter. My hopes of returning the new rig were dashed when I found out Hallicrafters had been out of business for several months. That experience taught me that it's best to read the *QST* Product Review before buying any new equipment.

Kenwood's TS-990S is an example of a sophisticated SDR in a conventional case with knobs and buttons.

The FlexRadio FLEX-3000 is a black-box SDR.

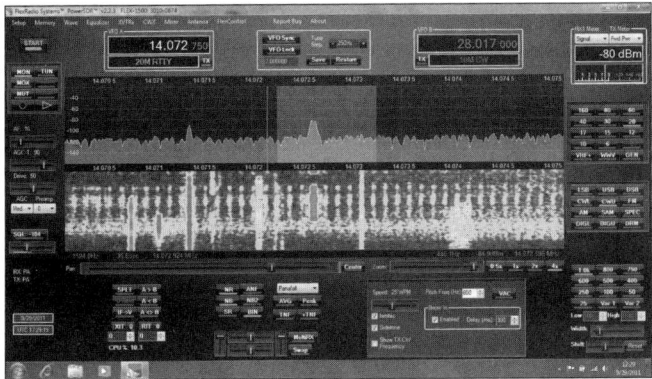

FlexRadio transceivers rely on *PowerSDR* software for the user interface.

function, feature and control on a computer monitor. A click or drag of the mouse or a keystroke does all the work, with esthetics supplied by spectral and waterfall displays that mesmerize the user.

Displays come in many shapes, sizes and colors. Illumination may be from a backlit liquid crystal display (LCD), a thin-film-transistor liquid crystal display (TFT) display, a fluorescent display, or with some kit transceivers, a light-emitting diode (LED) display.

TFT displays, which use the same technology as most flat screen monitors, have been features of high-end transceivers for several years and are now being offered with transceivers in the mid-price range. The key advantage of a TFT display is that it allows many of the transceiver's operating parameters to be viewed at a glance. Some TFT displays include a panadapter, which was discussed in Chapter 12. This provides a spectral display of a portion of the ham band above and below the tuned frequency and is very handy for finding activity. The disadvantage of TFT displays is that they are fragile due to their thin surface, and care must be taken when transporting a transceiver with a TFT display.

LCD displays are very common on low- and mid-priced transceivers. Most have the option of adjusting brightness and contrast to suit the operating environment, and some offer a choice of background colors. I like a sky blue background, for instance; it's pleasing to look at and lessens fatigue during

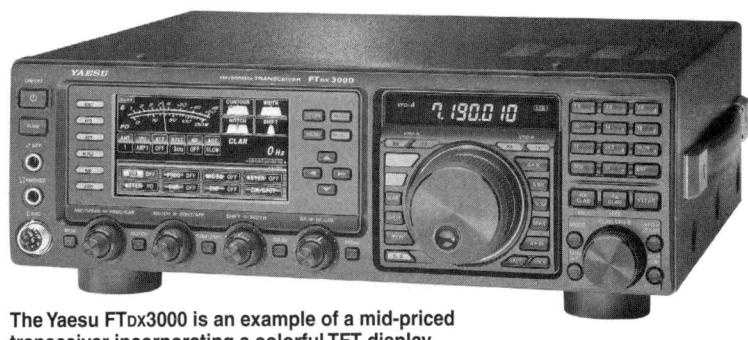

The Yaesu FTdx3000 is an example of a mid-priced transceiver incorporating a colorful TFT display.

Other Considerations   **20-3**

The simple LCD on Kenwood's TS-590S includes bar graph metering functions for transmit and receive, as well as a number of indicators showing the status of various transceiver settings.

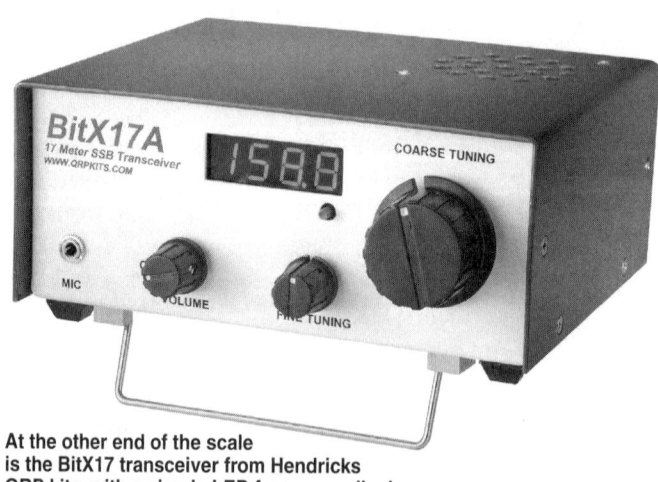

At the other end of the scale is the BitX17 transceiver from Hendricks QRP kits, with a simple LED frequency display.

long operating events. One disadvantage of LCDs is their limited viewing angle. The display is clear when viewed directly from the front, but it will look fuzzy or even disappear when viewed at an angle (the same is true of TFT displays). I have also seen some LCDs turn black — permanently — after the transceiver was stored in a very cold environment. I would not want to leave my transceiver in the car on a 0° night.

Icom's IC-7800 is a top-tier transceiver that features an elaborate TFT display, including a pair of simulated analog S meters.

S meters have provided our reference point for incoming signal strengths since the late 1930s. On most transceivers, the metering provides not only signal strength, but also a variety of transmitter information such as power output, SWR, supply voltage/current, ALC and speech compression level. The seasoned radio amateur is familiar with the analog meter and its associated mechanical pointer and movement. This type of meter is still seen, mostly on mid- and high-priced transceivers, but today metering is often incorporated in the LCD as a bar graph, or as a recreation of an analog meter on a TFT display.

A full SDR transceiver has receive metering in both S units and dBm, and while an S unit is handy at a glance, the operator has to do a calculation to convert the incoming signal strength to dBm. The operator of an SDR transceiver can read both meters and see the signal strength, in dBm, on the spectral display itself. All meters do the same basic job, and it's a matter of personal preference to the purchaser.

20.3 The Feel

The feel of a transceiver is difficult to assess by looking at a printed advertisement, yet it greatly influences the level of operating satisfaction, so the author of each *QST* Product Review tries to express his or her opinion about "the feel." Probably the best example of feel is provided by the tuning knob. A weighted, metal tuning knob with a rubber grip just "feels" better than an inexpensive plastic knob of the same size. Control knobs that have excessive play in the shafts they're mounted on can feel less than solid and be annoying. The same goes with buttons and switches.

20.4 The Smell

Like the latest model car, a new transceiver has that "new smell" to it. The purchaser of a new amateur product gets to enjoy that same that sensation. If you're going to buy a used transceiver, however, always ask if it was used or stored in an environment where someone was smoking. The tar exhaled by a smoker not only degrades the look and the performance of a transceiver, but also clings to the device and will make the shack of a non-smoker smell of stale tobacco. The seller may be able to clean up the transceiver on the outside, but that won't help with the smell once the transceiver is powered up.

20.5 The Sound

As with high-fidelity audio equipment, speaker size and placement on an amateur transceiver is important for listening pleasure. For best audio quality it is always preferable to have a good-size, front-facing speaker, but this is nearly impossible these days because of the sheer number of controls and displays on a modern transceiver. A mobile transceiver sitting on a table with a speaker on the bottom cover is not ideal. Fortunately, manufacturers realize this and provide an external speaker output jack. Many makes and models offer an optional matching external speaker or station monitor that can dramatically improve listening pleasure and reduce fatigue during long operating events.

Audio filters adjusted for a narrow frequency range to improve selectivity, or adjusted for a wider range to provide fidelity, may enhance your experience. Conversely, some SDR transceivers can widen out to an audio bandwidth of 10 or 15 kHz, which is very useful for listening to traditional AM broadcast and shortwave stations. Higher audio frequencies (up to 5 kHz, sometimes higher) are actually transmitted by the broadcaster, but are not heard with many traditional transceivers that have a high frequency roll-off due to the frequency conversion process and filtering. AM can sound pretty good, too, through the proper receiver and speaker.

20.6 The Taste

No, we don't expect you to take a bite from your equipment, but we have to acknowledge that we all have our individual tastes when it comes to transceivers and other station equipment. The likes and needs of one amateur may not be the same as those of another.

Casual operators, working with simple wire or vertical antennas, can be perfectly happy using a small mobile transceiver in their home shack, and it may perform just as well as a more expensive transceiver designed to

The Alinco DX-SR8T is an inexpensive 160-10 meter transceiver. Its front panel is detachable for easier installation in a mobile station.

sit on a desk. There is one exception, however. If the amateur has enough room for a full-sized 160 or 80 meter dipole or loop antenna, he or she can experience extremely strong signals that can cause undesired effects during crowded conditions. Sometimes high signal levels, well above 40 dB over S9, are possible via sky wave. If you plan to operate with a full-sized low-band antenna, consider spending a little more money for a transceiver with higher dynamic ranges.

The mobile operator usually will not experience very strong signals (except from radio amateurs in passing vehicles) and should be more concerned with the ability of the transceiver's noise blanker or digital noise reduction circuits to reduce noise from manmade sources such as power lines, automotive ignition systems, and lighting. The ruggedness of the transceiver and the microphone is important as well. Some mobile transceivers can interface directly with a matching mobile antenna (and/or automatic antenna tuner), which automatically adjusts for the lowest SWR. Detachable control panels are available on some HF mobile transceivers, facilitating installation.

Of course, the small size of portable and mobile transceivers means there's less space for knobs and buttons on the front panel. Luckily, having fewer knobs and buttons does not translate into reduced functionality. Fewer buttons usually means more menus. Menu functions and controls are usually configured before setting off on a trip, and once underway the operator should only need to make minimal adjustments to the mobile transceiver.

The portable operator (hiker, backpacker, public service volunteer) is mostly concerned with current consumption, since lightweight batteries or solar panels are used as the power source. Typically, with a good antenna

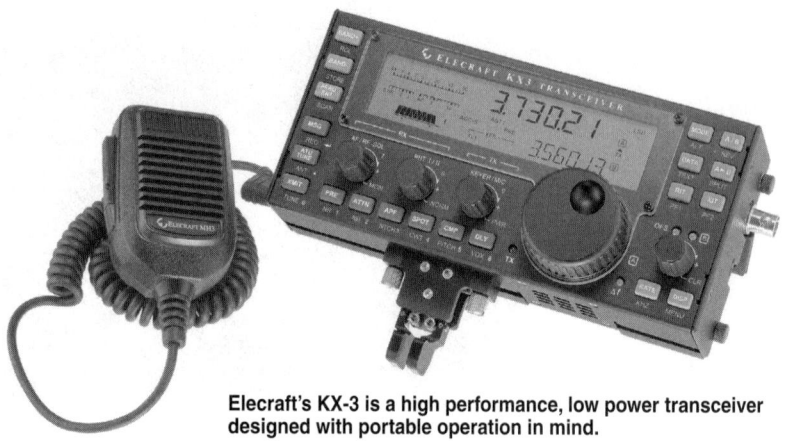

Elecraft's KX-3 is a high performance, low power transceiver designed with portable operation in mind.

and favorable propagation, a portable operator can communicate effectively with 5 W RF output or less. Battery life can be extended by carefully adjusting the power levels to a minimum. Knowing the capacity of the battery (in amp-hours) and the current consumption in both transmit and receive mode, the portable operator can calculate the expected time he or she can remain on the air. Something as simple as turning off the backlights of the display and keeping the volume low will help conserve energy. Small mobile transceivers, capable of 100 W RF output, also can be used for portable operation if the RF power output is reduced to a minimum. For such transceivers, check for the lowest possible RF power output available; some transceivers can go only as low as 5 W, but lower is preferred.

In additional to performance, serious DXers and contesters want comfort, ease and flexibility while operating their transceivers. This is where ergonomics become important. The operating position itself must be comfortable if substantial time will be spent at that location. The proper chair with the right support, the height of the desk, and the comfortable placement of the wrists are all important factors in reducing fatigue. A transceiver should be located on top of a desk or table, not on a shelf where you have to reach up to tune the dial, turn knobs, or press buttons.

The layout of a transceiver can reduce or cause fatigue. Until the operator has memorized all the buttons, knobs and switches that control the various functions, he or she must rely on the labeling. This may be daunting for some mid- and high-priced transceivers, not only because of the sheer number of controls, but also because of how each control is labeled. A black panel face with light labeling, or a white panel with black labeling is easiest to read. The label is also easier to read if it is located above the control. Labels located

The Icom IC-9100 covers 160 through 2 meters, plus 70 cm with an option for 23 cm (1.2 GHz) operation. It uses a large, highly readable LCD screen to clearly display operating frequency and other information.

below a control can be hard to see, especially if the transceiver is at desktop level, as the user must scrunch down a bit to read them.

Some mid- and high-priced transceivers offer illuminated buttons that control key functions. Other options to let the user know if a function is engaged include LED lights or icons displayed on the screen. I prefer a separate light or LED indicator, since icons on the main display can be easy to miss. However, that illuminated button or LED indicator makes it more expensive for the manufacturer to produce so the purchase price goes up.

Regardless of the display type, indications of frequency, signal strength, and ALC metering must be clearly readable from at least arm's length. The complaint heard most often about a transceiver's display is that the numbers are too small. I would agree, but in order for a transceiver to be small enough for mobile or portable use, the display must also be small so the front panel can accommodate other controls. When shopping around for a transceiver, consider lighting and the likely location of the device to make sure the display is readable at the operating position.

20.7 The Support

Should you opt for a new or used transceiver? Before making that decision, consider if the transceiver will be repairable if it breaks. New transceivers usually come with a warranty of up to a year. Personally, I run a new transceiver as much as possible during its first year, figuring that any failure will occur during this time and I'll be covered for it. After the transceiver's warranty period has expired, the manufacturer is expected to

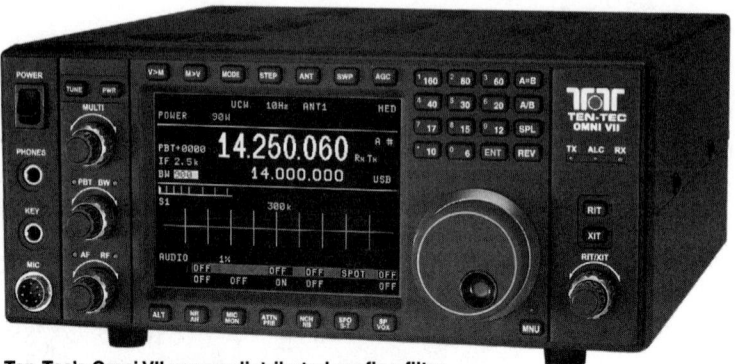

Ten-Tec's Omni VII uses a distributed roofing filter architecture to improve performance. With manufacturing, sales and service in Tennessee, Ten-Tec has a long-standing reputation for excellent customer support.

be able to service its brand for a number of years, even after production of a model is halted, until the stock of parts runs out. There are also independent businesses that repair Amateur Radio equipment, provided they can get the parts. Many key components, such as microprocessors, displays and other integrated circuits, are manufactured for a specific purpose in a specific transceiver and are not easy to come by. There are no manufacturers who are willing to set up and run a production line for a small number of obsolete parts, so if you want to purchase a used transceiver, find out if it still serviceable and, if so, for how long.

I have a life-long friend who needed a used transceiver to get on the air after a long hiatus from Amateur Radio. With good intentions, I purchased an older used HF transceiver for him at a good price. All was well for a few weeks, until the transceiver's display went haywire and the radio failed to operate. The manufacturer wouldn't touch it since it had been out of production for almost 15 years. I found an independent service shop to look at it, only to learn that the parts were no longer available to repair the transceiver and it was now a glorified doorstop. My options were to find a "junker" with good parts in it or send it to the recycling center.

With today's rapidly changing technology, it is wise to purchase new, or at least to purchase an old transceiver that you can still find parts for. I own modern equipment, but I also have a Heathkit HW-101 transceiver that uses discrete components and vacuum tubes. Thousands of these were built and there are plenty of HW-101 parts around today. The vacuum tubes and other components can still be found easily. However quaint the technology of the HW-101, it lacks the performance and the amenities of a modern transceiver.

20.8 The Price

I'm purposely ending this book with the topic that most radio amateurs think of first when choosing an HF transceiver: the price. I hope you're an exception and have considered your operating needs first.

The price of a new HF transceiver can be intimidating. It always has been. If you take inflation into consideration, today's amateur equipment is less expensive and offers more value than ever before. For example, let's look at the Kenwood TS-520S, an oldie but goodie from 1978. Back then, $649 got you this transceiver, which covered the 160, 80, 40, 20, 15 and 10 meter ham bands only. It offered two modes, SSB and CW. Extra filtering for CW was optional. The purchase price in today's dollars is over $2400. For slightly more than one-third of that price, you can now purchase a small 100 W transceiver with 160, 80, 60, 40, 30, 20, 17, 15, 12 10, 6, 2 meter and 70 cm transmitting capabilities, general coverage up to 470 MHz, SSB, CW, AM, FM and digital modes, and DSP technology.

It's often difficult to raise the cash for a new HF transceiver, but when the funds are finally available for the purchase of your transceiver, whether new or used, I do hope you'll take the time to read the ARRL's *QST* Product Review for the transceiver you're interested in. I also hope this book has helped in the decision-making process. Let me leave you with one last piece of advice: If your search is rooted solely in the lowest possible cost of the transceiver, you are missing the big picture and will likely be disappointed with your purchase, one way or another.

The classic Kenwood TS-520S from 1978 sold for about $2400 in today's dollars and lacked many of the features we take for granted in a modern HF transceiver.

Glossary

Attenuator — A broadband device that reduces the amplitude of a signal by a specified, well-controlled amount.

Blocking gain compression dynamic range — The difference between the *blocking level* and the *MDS*.

Blocking level — The level of an interfering signal that causes weak signals to be reduced in amplitude by 1 dB.

Dynamic range — The difference in dB between the strongest and weakest signals that a receiver can handle.

Harmonic signal — A periodic signal. Its frequency spectrum consists only of the fundamental and harmonics of the repetition rate.

Hybrid combiner — A passive device used to combine two signals in such a way as to reduce the interaction between the signal sources.

IF and image rejection — The difference in level between an interfering signal at a receiver's IF or image frequency and the level of the desired signal that produces the same response.

IMD (intermodulation distortion) — The creation of unwanted frequencies because of two or more strong signals modulating each other.

IMD dynamic range — The difference in signal level between two signals that cause IMD products at the level of the MDS and the MDS.

MDS (minimum discernible signal) — The level of the *noise floor* at the antenna connector of a receiver. Depends on the measurement bandwidth.

Noise figure — A figure of merit for the sensitivity of receivers and other RF devices. It is the ratio of the effective noise level to the thermal noise level, usually expressed in dB.

Noise floor — The noise received by a receiver in a specified bandwidth, referenced to the antenna connector. It is the thermal noise in dBm plus the noise figure in dB.

Oscilloscope — An instrument that displays signals in the *time domain*. Has a graphical display that shows amplitude on the vertical axis and time on the horizontal axis.

Reciprocal mixing — The mixing of a nearby interfering signal with the phase noise of the receiver local oscillator, which causes noise in the audio output.

Resolution bandwidth — The smallest frequency separation between two RF signals that a *spectrum analyzer* can resolve. It is determined by the IF filters in the *spectrum analyzer*.

Second-order IMD — IMD caused by strong signals at frequencies f1 and f2 that occurs at a frequency of $f1 + f2$.

Signal generator — An instrument that generates a calibrated variable-frequency RF signal, usually with adjustable amplitude and modulation capability.

SINAD (signal plus noise and distortion) — A measure of the relative level of signal compared to the noise and distortion at the audio output of an FM receiver.

Spectrum analyzer — An instrument that displays signals in the *frequency domain*. Has a graphical display that shows amplitude (normally in logarithmic, or dB, form) on the vertical axis and frequency on the horizontal axis.

Spurious emissions, or spurs — Unwanted energy generated by a transmitter or other circuit. These emissions include, but are not limited to, harmonics.

Step attenuator — An attenuator that can be switched between different attenuation values.

Third-order IMD — IMD caused by strong signals at frequencies f1 and f2 that appears at frequencies $2f1 - f2$ or $2f2 - f1$.

Third-order IMD intercept point (IP3) — The power level representing the intersection of plots on a dBm scale of the power level of the two strong signals and their IMD products.